Addressing the Friction Between the Army's People First Initiatives and Its Readiness Generation Process

MICHAEL E. LINICK, JEREMY M. ECKHAUSE,
LISA SAUM-MANNING, BRUCE R. ORVIS, ANDREA M. ABLER,
SAM WALLACE, PHOEBE FELICIA PHAM, SARAH BAKER

Prepared for the United States Army
Approved for public release; distribution unlimited

ARROYO CENTER

For more information on this publication, visit **www.rand.org/t/RRA2006-1**.

About RAND

The RAND Corporation is a research organization that develops solutions to public policy challenges to help make communities throughout the world safer and more secure, healthier and more prosperous. RAND is nonprofit, nonpartisan, and committed to the public interest. To learn more about RAND, visit www.rand.org.

Research Integrity

Our mission to help improve policy and decisionmaking through research and analysis is enabled through our core values of quality and objectivity and our unwavering commitment to the highest level of integrity and ethical behavior. To help ensure our research and analysis are rigorous, objective, and nonpartisan, we subject our research publications to a robust and exacting quality-assurance process; avoid both the appearance and reality of financial and other conflicts of interest through staff training, project screening, and a policy of mandatory disclosure; and pursue transparency in our research engagements through our commitment to the open publication of our research findings and recommendations, disclosure of the source of funding of published research, and policies to ensure intellectual independence. For more information, visit www.rand.org/about/research-integrity.

RAND's publications do not necessarily reflect the opinions of its research clients and sponsors.

Published by the RAND Corporation, Santa Monica, Calif.
© 2023 RAND Corporation
RAND® is a registered trademark.

Library of Congress Cataloging-in-Publication Data is available for this publication.
ISBN: 978-1-9774-1266-9

Cover: U.S. Army photo by Spc. Dana Clarke, Operations Group, National Training Center.

About This Report

This report documents research and analysis conducted as part of a project titled *Enabling "People First" Through Army Synchronized Readiness*, sponsored by U.S. Army Forces Command (FORSCOM). The purpose of the project was to evaluate the risks to the Army's People First initiatives that result from competing priorities for time and resources required to achieve manning and readiness goals across demands on Army operating force capabilities ranging from peacetime missions to the ability to deter and execute primary wartime contingency plans.

This research was conducted within RAND Arroyo Center's Personnel, Training, and Health Program. RAND Arroyo Center, part of the RAND Corporation, is a federally funded research and development center (FFRDC) sponsored by the United States Army.

RAND operates under a "Federal-Wide Assurance" (FWA00003425) and complies with the *Code of Federal Regulations for the Protection of Human Subjects Under United States Law* (45 CFR 46), also known as "the Common Rule," as well as with the implementation guidance set forth in U.S. Department of Defense (DoD) Instruction 3216.02. As applicable, this compliance includes reviews and approvals by RAND's Institutional Review Board (the Human Subjects Protection Committee) and by the U.S. Army. The views of sources utilized in this report are solely their own and do not represent the official policy or position of DoD or the U.S. government.

Acknowledgments

We would like to thank Kirk Palan and Dana Luton for exceptional support to this project. They provided excellent guidance on FORSCOM concerns and helped us immensely in gaining access to Army personnel and organizations. We would like to thank the Army officers, noncommissioned officers, and civilians who participated in our interviews and provided us with documents necessary for our research. Our RAND colleagues, Christine DeMartini, Candice Miller, and Rodger Madison, were invaluable in helping us work through data access, cleaning, merging, and analysis. We would also like thank our editor, Nora Spiering, for greatly improving the appearance and clarity of this document. Finally, we thank our reviewers, John Ferrari and Peter Schirmer, for their very helpful comments and suggestions that greatly improved the quality of this report.

Summary

Introduction

Prompted, in part, by the results of the Fort Hood Independent Review Committee's Report, the Army has adopted and promoted a set of initiatives collectively referred to as "People First."[1] People First began as a discrete set of programs designed to help the Army confront unhealthy behavior in the ranks (e.g., sexual assault or harassment and violent extremism), build cohesion and understanding in the ranks (including diversity, inclusion, and equity), manage Army talent differently, and provide support to soldiers and their families in dealing with unique aspects of Army life.[2] Simultaneously, the Army has designed a new model for how it wants to produce ready units over time: the Regionally Aligned Readiness and Modernization Model (ReARMM). As the name implies, ReARMM includes specific time frames dedicated to the modernization of units and provides units with a regional focus to help guide their manning, equipping, and training, as well as, ultimately, the missions for which they will deploy or be prepared to deploy.

The challenge for the Army is that, in some ways, the imperatives of the People First initiatives and programs are at odds with the successful achievement of units' ReARMM goals and milestones. Soldiers have a limited amount of time to spend on either set of goals, and commanders must decide how to allow soldiers to spend that time. It is this friction between People First and ReARMM that this study examines.

---

[1] The Fort Hood Independent Review Committee was appointed by the Secretary of the Army to "conduct a comprehensive assessment of the Fort Hood command climate and culture, and its impact, if any, on the safety, welfare and readiness of our Soldiers and units" Christopher Swecker, Jonathan Harmon, Carrie Ricci, Queta Rodriquez, and Jack White, *Report of the Fort Hood Independent Review Committee*, U.S. Army, November 6, 2020, p. ii).

[2] U.S. Army, "Army People First: Prioritizing Our Most Valuable Asset—People First Task Force," webpage, undated.

Army leaders must make decisions daily about how to balance their mission requirements with supporting the needs of soldiers, many of which do not directly contribute to near-term unit readiness. Leaders also recognize that the Army is in competition with the civilian sector in a national crunch for talent. Striking the right balance between maintaining unit readiness and the work-life balance and professional development interests of soldiers will have an impact on how the Army recruits and retains its strength. Overcoming such challenges embodies the friction between ReARMM and People First. Our research revealed that, for each primary focus area in ReARMM, the decision criteria that commanders employ may vary, being more "people friendly" in some phases and more "mission focused" in others.

The Army's personnel system also impacts this process. The Army's talent management initiatives, a key part of the People First initiatives, focus on longer-term readiness issues: fully developing the capabilities of soldiers by effectively matching their knowledge, skills, abilities, and desires with job assignments that are compatible.[3] One key aspect of this is the creation of a talent marketplace to replace the old, centralized assignment system. In this marketplace, soldiers are given increased (but not absolute) autonomy over their assignments.[4] If talent management is done well, it is believed it will lead to more effective soldiers and to better retention rates. But this additional level of autonomy in job selection adds to the challenges of manning the force in a way that balances soldier desires with unit needs. The effect of this marketplace is that it is not uncommon for units to be missing key personnel at different points in their unit's ReARMM cycle or to have a limited number of those personnel available, which can impact how commanders think about the work-life balance of those specific soldiers.

---

[3] For information about the Army's talent management program, see both the Army's People Strategy (U.S. Army, *The Army People Strategy*, October 2019b) and the Army's talent management website and accompanying literature (U.S. Army Talent Management, homepage, undated-a).

[4] This is through a program called the Army Talent Alignment Process. See, for example, U.S. Army Talent Management, "Army Talent Alignment Process," webpage, undated-b.

Another key Army goal is the development and maintenance of cohesive teams. General James McConville (Chief of Staff of the Army [CSA] at the time of writing), said:

> There are three areas that I am most concerned about that are breaking trust with the American people and hurting our Soldiers: sexual misconduct, suicides, and racism. These are what building cohesive teams and the Sergeant Major of the Army's "This is My Squad" initiatives get after. . . . It's about having tough conversations to make sure that leaders understand everyone's perspective. More importantly, it's about getting to know each other's story and have a deep understanding and appreciation for each other.[5]

In short, commanders must balance soldier needs for work-life balance, personal and professional development, and time with family, as well as time for leaders to build cohesive teams with their soldiers, against the units' needs for soldier time and effort focused on unit mission training. The institution also must balance having the right soldiers assigned for commanders to employ with other demands for soldiers' time and preferences. Thus, friction is caused by the need to balance whether the correct soldier is available (an institutional issue) and, given the soldier's availability, the decisions about how to employ that soldier on a day-to-day basis. The key research question that U.S. Army Forces Command (FORSCOM) asked the RAND Arroyo Center to explore was as follows: Are there ways to further reduce the challenges associated with managing this balancing act between mission (broadly defined) and work-life balance, especially given the Army's senior leadership focus on putting people first?

Methodology

Our research relied on six distinct levels or methods of research. First, after discussing the research with our sponsor and identifying key policy documents and references, we conducted a thorough policy review and refined

---

[5] James C. McConville, "People First: Insights from the Army's Chief of Staff," *Army Sustainment*, Vol. 53, No. 1, 2021, p. 21.

our understanding of our research questions and approach. Following the policy review, we conducted interviews at the policy level and at the unit level to understand the macro-level policies that intersected with ReARMM and People First. Our goal with the policy-level interviews was to determine to what extent, if any, we could reveal synchronization problems and explore where the Army might have the ability to adjust them, if necessary and possible—a policymakers perspective. The two specific areas of focus for synchronization questions involved the Army's system for assigning and reassigning personnel—the Personnel System—and the Army's Modernization System, responsible for delivering new equipment to soldiers and training them on how to use and maintain it.

Our unit-level interviews were focused on understanding how People First initiatives and ReARMM were perceived and what friction points had been identified at the corps, division, and brigade levels—a policy-"takers" perspective. As a result of this, we also identified, and subsequently gathered, additional lower-level policy and prioritization documents (e.g., corps' commanders' training guidance). Our third research approach was to take the accumulated interview and policy work and apply a lexical analysis tool. This helped us systemize and quantify key themes that emerged from the interviews and document reviews.

As we began to understand the key themes, and specifically as our interviews and document reviews started to point to the role of culture in creating and managing (or not managing) friction, as a fourth approach, we decided to conduct a small experiment involving a convenience sample of Army field-grade officers participating in a role-playing workshop. This scenario-based workshop, while not conclusive from a statistical standpoint, helped us confirm some findings and further refine our recommendations.

We wanted to take anecdotal evidence we had gleaned from our interviews and see whether those experiences were supported by data analysis. Thus, our fifth approach was to use Army personnel data and examine them in conjunction with ReARMM cycle calendars provided by FORSCOM. Our interviews did not indicate that there was any evidence of widespread scheduling problems for requirements and resources. Even if the Army would consider major ReARMM cycle changes in response to People First concerns, these large, generalizable events appeared to be well managed. As a result, traditional prescriptive analytics of these systems (e.g., optimal schedules) were unlikely to be either necessary or implementable. We instead identified descriptive and predictive analyses that would allow us to develop descriptive statistics about soldiers' experiences and how those intersect with their unit's ReARMM cycle, which then inform recommendations for improving the friction. We identified seven areas of interest for exploration and were able to gather and analyze data to analyze most of them.

Our final approach was to take a limited look at literature on expectation management and see whether there were insights from that literature that might be applicable to some of the Army's ReARMM and People First friction challenges. Given uncertainty in unit schedules (despite ReARMM's best intentions) we wanted to see whether we could derive recommendations on how to best communicate and manage uncertainty.

Findings

Our research revealed that the macro systems that interface with ReARMM—the Army's personnel and modernization processes—have little room and little incentive to change their basic processes to synchronize better with ReARMM or to enable ReARMM to better accommodate People First goals. However, we also found that this might not be necessary and that a focus on culture changes and improved communications might be more productive.

Systemic Challenges Create Friction Points Between ReARMM and People First Goals

Our interviews suggest (and other research has supported the idea) that the challenges facing Army units are as much cultural as they are systematic.[6]

---

[6] See, for example, Lisa Saum-Manning, Tracy C. Krueger, Matthew W. Lewis, Erin N. Leidy, Tetsuhiro Yamada, Rick Eden, Andrew Lewis, Ada L. Cotto, Ryan Haberman, Robert Dion, Jr., Stacy L. Moore, Michael Shurkin, and Michael Lerario, *Reducing the Time Burdens of Army Company Leaders*, RAND Corporation, RR-2979-A, 2019. Also, see Todd C. Helmus, S. Rebecca Zimmerman, Marek N. Posard, Jasmine L. Wheeler, Cordaye Ogletree, Quinton Stroud, and Margaret C. Harrell, *Life as a Private: A Study*

The Army's mission is inherently unpredictable—so systems designed to produce predictability will, almost by definition, occasionally fail to do so. Even though the Army personnel we talked to had various interpretations of what People First means and requires within the context of ReARMM's different phases, there was general consensus that the real goal is to have the right people, at the right time, with the right equipment, at the right event—whether that event was a real-world mission, training, or other unit activity—while still providing predictability, stability, and work-life balance. The challenges of doing so reflect challenges of both availability and employment of soldiers. By availability, we mean, is the correct soldier, with the correct skills, assigned to the unit at the time of need? And, by employment, we mean that even when the correct soldier is assigned to the unit, is that soldier available for the unit to use, or to use in ways that promote unit-level readiness?

A Lack of Consistent Guidance Leaves Room for Various Interpretations of What to Prioritize

This leader's dilemma leads to another major finding of the research. Although Army senior leaders, from the CSA down, all provide guidance on how to think through decisions about what to tell a soldier about time off or other activities that are not directly related to a unit's readiness report, the message is not being evenly heard or understood at lower levels. One reason this occurs is because the messaging is diluted as it moves down the chain of command.

We examined policy from the CSA level, through the FORSCOM commander's guidance and corps commanders' guidance for FORSCOM-assigned corps headquarters, and down to division commanders' guidance for some of the FORSCOM assigned divisions. What we found was that, at the top, the CSA and FORSCOM commander articulated a clear set of priorities, focused on people and acceptant of readiness risk. By the time this

of the Motivations and Experiences of Junior Enlisted Personnel in the U.S. Army, RAND Corporation, RR-2252-A, 2018; and James V. Marrone, S. Rebecca Zimmerman, Louay Constant, Marek N. Posard, Katherine L. Kidder, Christina Panis, and Rebecca Jensen, *Organizational and Cultural Causes of Army First-Term Attrition*, RAND Corporation, RR-A666-1, 2021.

guidance had been translated down to units, the overwhelming focus of all guidance was on gaining and maintaining unit readiness: The People First priorities had disappeared or had been so subsumed into readiness language that it was clear that the focus was on training, not people.

Lack of Compelling Incentives to Implement People First Goals

In addition to inconsistent messaging, the units perceived that there were no discernable rewards for doing well at People First. There were, however, clear incentives to achieve training goals (and disincentives to failing to achieve them), and the success criteria for those readiness goals were essentially the same for each ReARMM phase the unit was in. Essentially, training and unit performance were directly observable and measurable and could be attributed to the specific chain of command under which the goals were met (or not). People First goals were not specified in a way that could be easily observed or measured, and the output of focusing on those goals could not be clearly attributed to a specific person or chain of command. If units do best only that which is measured, inspected, or reported, then the current system is focused not on People First, but on training outcomes.

Training Management Lacks Effective Synchronization with ReARMM and People First Goals

Several people we interviewed noted that applying ReARMM and People First goals may not be easily integrated with the Army's systems for training management. Our discussions with both institutional offices and with unit leadership surfaced synchronization challenges, such as having a lead time to request or coordinate training resources (e.g., ranges, ammunition) that is longer than the planning horizon for the unit—leading to missed or poor training opportunities. Interviews revealed mixed views on whether People First principles are applicable to the combat training center context, whose core mission set places a premium on unit readiness over what might be considered People First–related individual needs of the soldier.

We also heard from several unit leaders, at more than one installation, that when company-level (and even battalion-level) units kept "white space" (i.e., space that did not have a dedicated training or mission purpose) on their calendars, higher-level headquarters interpreted that as excess capacity and wanted to fill that white space with additional taskings or requirements.[7] In contrast, the units looked at that space as time when they would have maximum flexibility to address some of the People First goals—including allowing soldiers time to take care of personal or family issues, to catch up on other mandatory training, or to focus on activities designed to build cohesion within the squads or teams.

ReARMM Is a Standardized Concept, While Unit Requirements Vary

Several interviewees clearly articulated that ReARMM seems to be a brigade combat team–centric (BCT-centric) model that does not fit well with units that either have limited modernization needs or have significant and ongoing missions (e.g., sustainment or intelligence units).[8] We note that FORSCOM has made clear that (as of mid-2022) ReARMM was still being developed and fielded and that FORSCOM was already preparing new policy that addressed adjusting ReARMM to better support these non-BCT units.[9]

Soldiers Rely on Other Soldiers First and Big Army Last When Faced with Work-Life Balance Challenges

We held a workshop with field-grade officers with a wide range of operational experience who provided critical insights not captured elsewhere in

---

[7] Sgt. Maj. of the Army Michael Grinston discusses this problem, indicating that the Army leadership recognizes the white-space problem. The anecdote provided indicates the cultural challenge that the Army has in addressing it. See Haley Britzky, "The Sgt. Maj. of the Army Wants Leaders to Stop Scheduling Training Just for the Sake of It," *Task & Purpose*, October 18, 2021.

[8] Our interview population will be fully described below, but note for this section that we did not interview below battalion level. Therefore, our evidence is based on what battalion and brigade leaders (as well as institutional staffs) observed and reported about company-level expertise and experiences.

[9] Further research would need to be conducted to see how well those new policies address the issues we identified (or the ones FORSCOM had already identified).

our research. Workshop participants served in combat support and combat service support positions, worked in garrison, had deployed experience, and collectively served in all three of the Army's components: Regular Army, National Guard, and Reserves. These officers shared personal experiences dealing with many of the kinds of friction dilemmas we explored in this study. Perhaps the biggest takeaway is that soldiers typically rely on their fellow soldiers first and "big Army" last when faced with making difficult trade-offs between work requirements and family obligations. In other words, the first recourse seems to be to work around the ReARMM system, rather than to reform it. While the buddy system is a critical resource for soldiers, implementing practical and feasible Army policies should be as well.

Measuring and Predicting Friction: Findings from Quantitative Analyses of Personnel and ReARMM Data

We analyzed major Army datasets to understand the degree of conflicts involving not having the right person at time of need and to see how personnel outcomes correlated with changes in personnel status. While friction is clearly happening, it is possible that the level of incidents may be less than commonly believed. Our quantitative analysis of personnel data revealed that the level of mismatch between soldiers' military occupational specialty (MOS) and their duty MOS is generally small but is statistically larger for the deployment phase of ReARMM for FORSCOM BCT units. This result appears to support some concern expressed in the interviews about mismatch. While assignment considerations for family or conduct appear to emerge slightly more frequently during the training phase of ReARMM, they are not significantly higher during deployment. Conducting this type of analysis for non-BCTs may yield results that are less intuitive and more insightful. Examining duty location preference reveals that overall preference to actual location match is low—about 25 percent for those that indicated a preference—and may reflect some degree of work-life balance concerns. We found some differences in resident course attendance over the ReARMM cycle, but the overall level of attendance was very small, and the differences across the cycle were even smaller.

We find that turnover in units through the ReARMM cycle paints a nuanced picture. The overall monthly gains and losses for units are generally in the Human Resources Command (HRC) aggregate 2–3 percent target range. In addition, there do not seem to be increased gains and losses during less desirable times, such as towards the end of a training phase. On the other hand, while the aggregate monthly gains or losses are roughly 2–3 percent, there are significant statistical outliers, with several in the 7–8 percent range. These results may be useful as FORSCOM explores how to address and manage change oriented toward minimizing friction.

While friction is clearly happening and was suggested by the interviews, it is possible the level of incidents (as suggested by the quantitative data) may be less than commonly believed. For most dimensions of the quantitative analysis, the data suggest that most of the time, for most units, the problems do not appear to be of a large magnitude.

Some People First metrics are associated with retention. A soldier with a third-degree MOS match (i.e., the soldier's MOS, Special Qualification Identifiers, and Additional Skill Identifiers all match their duty assignment) was slightly more likely to be retained. Several assignment considerations were also associated with retention: Soldiers with conduct considerations were much less likely to be retained than soldiers without those considerations. Soldiers with medical considerations also were much less likely to be retained. Women with family considerations were considerably less likely to be retained.

Recommendations

Our interviews, discussions, literature and policy review, and quantitative analysis suggest a variety of approaches to mitigating the friction at unit level. Our key recommendations are as follows.

Establish Clear, Consistent Prioritization Guidance

One key recommendation we make is that FORSCOM be as specific as possible on the readiness trades that it (or, more properly, its commanders) is willing to take in support of quality of life and work-life balance issues and in support of talent management issues—for example, stating that weapons qualification or crew qualification metrics in the modernization phase can be at a lower percent than otherwise required. This direction, once documented in command training or readiness guidance, should be used by subordinate headquarters as those headquarters develop their own derivative guidance. We recommend that FORSCOM review and require edits to subordinate corps and division readiness and training guidance to ensure that those documents are nested with the FORSCOM guidance.[10] Corps and division commanders should do the same, so that readiness guidance all the way down to the battalion level can be confirmed as clearly being consistent with FORSCOM's original intent.[11] Command-level leaders should consider developing and disseminating clear examples of how to execute that intent in recognition that leaders' interpretations of guidance will vary. While it may be very difficult to establish clear and consistent guidelines, given the amount of gray area surrounding people-versus-training issues, having this discussion as a point of emphasis at command training briefings (where a battalion briefs their quarterly or annual training plan to brigade leadership, or a brigade does so to division leadership, etc.) may help give commanders more clarity and understanding about expectations and decisionmaking.

Establish Indicators to Measure Progress Toward People First and Incorporate Them into Evaluation Mechanisms

We next recommend that the Army look at ways to include People First input into leaders' efficiency reports. To do this, the Army will have to think

---

[10] Army doctrine directs that "[c]ommanders train and resource training one echelon down, and they evaluate to two echelons down" (Field Manual 7-0, *Training*, Headquarters, Department of the Army, June 2021, p. 102, para. 1 6). See also this quote: "Soldiers two echelons down should easily remember and clearly understand the commander's intent. Commanders collaborate with subordinates to ensure they understand the commander's intent. Subordinates who understand the commander's intent are far more likely to exercise disciplined initiative in unexpected situations" (Army Doctrine Publication 6-0, *Mission Command, Command and Control of Army Forces*, Headquarters, Department of the Army, July 2019, p. 1-10, para. 1-50).

[11] By extension, we suggest that all troop-owning commands do the same. Specifically, U.S. Army Europe and Africa and U.S. Army Pacific also have assigned forces that would benefit from this type of additional oversight.

through how to quantify and measure, or at least to record and report, activities that fall under the rubric of People First. It very well may be that these manifest as required qualitative comments on a report—for example, "CPT Jones was able to maintain an effective work balance in the company by using effective training management skills that ensured every soldier had at least four hours of duty day per week to address personal or family issues and needs."

We stipulate that quantifying People First is problematic. At a macro level, People First is about reduction of sexual assault or harassment, reduction of suicide, and provision of talent management—managing and shaping the behavior and experiences of soldiers in ways that build cohesive teams. However, many of these behaviors and experiences cannot be easily quantified. Merely quantifying the time spent talking to or training soldiers about these issues is also not a good metric. Executing these tasks daily as part of unit culture and routine is basic Army leadership.[12] Over time, company- or battalion-level climate surveys may be able to identify the trends within a unit.[13] But, as tools to support officer or noncommissioned officer efficiency reporting, these assessment instruments may be lacking. Further expanding the Army's existing command climate assessment instrument to incorporate a wider range of People First issues may be something for the Army to consider. However, it will need to weigh or measure the opportunities to collect more data with the soldier and unit time costs of administering the surveys.

Manage Select Duty Positions Differently

The Army already manages some key duty positions more than it does the vast majority of its billets. One example of these intensively managed groups

---

[12] Both Field Manual 6-22, *Developing Leaders*, Headquarters, Department of the Army, November 2022, and Army Doctrine Publication 6-22, *Army Leadership and the Profession*, Headquarters, Department of the Army, July 2019, discuss the role of leaders in developing cohesive teams—including the need to have these kinds of discussions and interactions with soldiers.

[13] Army Regulation 600-20 contains the guidance on Army Command Climate Assessments. See Army Regulation 600-20, *Personnel—General Army Command Policy*, Headquarters, Department of the Army, July 24, 2020, Appendix E, pp. 139–145.

includes positions defined as "key and developmental." There may be opportunities to develop a similar set of key billets that specifically relate to identified friction points inside of ReARMM. For example, one of these opportunities may be to consider how to manage supply personnel differently and in ways that maximize the likelihood of having an appropriately trained supply specialist on hand when the unit is going through major equipment fielding. This type of focus may be manageable by both HRC and the personnel management chain in ways that a more generalized approach is not.

Leverage Personnel Data to Help Quantify and Manage Friction

Our quantitative analysis was done at the unit aggregate level but demonstrates an ability to do detailed analysis at grade-MOS level. Expanding the use of more-discrete analysis—like that which we present—may help the Army develop better policies that can alleviate some of the more critical disparities between needing a particular kind of soldier at a particular time and not having that soldier assigned to the unit at that time. One other aspect of that quantitative analysis, however, suggests that while it is true that there are occasions of missing the critical soldier (from an assigned standpoint), most of the cases suggest that those soldiers are present. Whether they are both present and employable in that critical role is, again, anecdotally reported but unmeasurable within the resources available to our study.

Assess Training and Execution Within the Army's Training Management System

Another recommendation concerns the Army's training management system. It was out of scope for this study to dive into the synchronization problems about which we heard, but we recommend that the Army do so. Many interviewees stated that training management expertise in the Army has been lost over time, so we recommend that the Army look at how it teaches training management and consider where it can improve. At a minimum, the Army should consider how to communicate or address the "white space" problem. Commander interviewees felt caught between the need to maintain the flexibility afforded by white space and the desire of higher headquarters to fill that space with additional guidance (usually in the form of taskings).

Continue to Tailor ReARMM to Accommodate More Variation by Brigade Type

As noted, interviewees clearly articulated a belief that ReARMM was an armored BCT–focused program. FORSCOM is developing policy that would address the differences between brigades and adjust ReARMM guidance to provide variations more closely tailored to the needs of other brigade types. We were not able to evaluate that new policy but recommend that the Army continue to develop a plan for evaluating the emerging policy.

Measuring and Predicting Friction: Recommendations from Quantitative Analyses of Personnel and ReARMM Data

We performed several dimensions of analysis to quantitatively assess metrics that appear in personnel databases with the ReARMM schedules. The analyses serve as examples of the types of inquiry that the Army can initiate to illuminate friction points between ReARMM and People First initiatives. First, these analyses investigated FORSCOM BCTs. To generate a clearer picture of forcewide implications of ReARMM, the Army should conduct similar analyses for other types of units as ReARMM continues to be implemented Army-wide. Because the implementation of ReARMM in BCT units is so recent, the Army should continue to conduct these types of analyses as ReARMM implementation becomes more mature. Specific analyses often tell opposing stories about friction points between ReARMM and People First initiatives. Some analyses suggest friction, while others—such as unit turnover—indicate little or no friction points at the aggregate level. However, the data also indicate that, despite predictability at the aggregate level, individual units, grades, MOS, or soldiers may experience outlier events that can be very disruptive. As a result, the Army should explore and balance the aggregate trends and the stress points for outliers along the distribution.

Table S.1 summarizes the findings and recommendations we discuss above. That said, there are no panaceas: Tensions between work-life bal-

TABLE S.1

Summary of Findings and Recommendations

Finding	Recommendation
• A lack of consistent guidance leaves room for various interpretations of what to prioritize.	• Establish clear, consistent prioritization guidance.
• There is a lack of compelling incentives to implement People First goals.	• Establish indicators to measure progress toward People First and incorporate them into evaluation mechanisms.
• Systemic challenges create friction points between ReARMM and People First goals.	• Leverage personnel data to help quantify and manage friction. • Manage select duty positions differently.
• Training and execution within the Army's training management system may need updates and improvements.	• Consider improvements to training on how to conduct training management. • Evaluate training management timelines to improve flexibility at the unit level.
• ReARMM is a standardized concept, while unit requirements vary.	• Monitor and observe new Army policies that adjust ReARMM policy for different brigade types.
• Soldiers rely on other soldiers first and big Army last when faced with work-life balance challenges.	• Employing the recommendations above will help the Army provide effective additional resources.

ance and mission are endemic to all work environments, and perhaps more pronounced in the military. Although development of people is critical, it also requires time not spent on mission-focused training. Balancing today's needs (mission) versus tomorrow's (personnel development, retention) creates dilemmas. Our research shows that Army systems may be adjusted to help address these issues, but only at the margins.

We also note that the full scale of the Army's current (2022–2023) personnel shortages began to manifest late into our research and was not apparent in the datasets we used. Consequently, our analysis does not consider the impacts on decisionmaking facing Army leaders as they grapple with the effects of those shortages and incorporate those challenges into their People First versus training and mission decision criteria. We can only surmise that it makes those decisions even more challenging.

We conclude by reiterating that for real change, both in the level of friction that soldiers and units experience and in managing the effect of those inevitable frictions, the Army must ensure that it communicates its priorities clearly all the way down to soldiers while aligning leader incentives with those priorities.

Contents

Figures and Tables

Figures

Tables

CHAPTER 1

Introduction

One way the Army has traditionally distinguished itself from other services is to note that while other services man equipment, the Army equips the man.[1] Soldiers are at the heart of the Army's view of itself, and taking care of soldiers is a part of its core leadership philosophy.[2] The Army's experience in recent years with the challenges of sexual harassment and assault, countering violent extremism in the ranks, and suicide have caused it to consider how it must adapt its personnel management and leadership to address these issues in the ranks. Prompted, in part, by the results of the Fort Hood Independent Review Committee's Report, the Army has adopted and promoted a set of initiatives collectively referred to as "People First."[3]

Simultaneously, the Army has designed a model for how it wants to produce ready units over time. The Army has evolved from the Army Force Generation (ARFORGEN) model, adopted in 2006 and designed to facili-

---

[1] See, for example, S. Rebecca Zimmerman, Kimberly Jackson, Natasha Lander, Colin Roberts, Dan Madden, and Rebeca Orrie, *Movement and Maneuver: Culture and the Competition for Influence Among the U.S. Military Services,* RAND Corporation, RR-2270-OSD, 2019, p. 49.

[2] See, for example, Army Doctrine Publication 6-22, *Army Leadership and the Profession,* Headquarters, Department of the Army, July 2019, p. 5-7, para. 5-39: "Taking care of subordinates is a solemn responsibility." Field Manual 6-22, *Developing Leaders,* Headquarters, Department of the Army, November 2022, contains similar language and emphasizes the role of leaders in developing cohesive teams.

[3] The Fort Hood Independent Review Committee was appointed by the Secretary of the Army to "conduct a comprehensive assessment of the Fort Hood command climate and culture, and its impact, if any, on the safety, welfare and readiness of our Soldiers and units" (Christopher Swecker, Jonathan Harmon, Carrie Ricci, Queta Rodriquez, and Jack White, *Report of the Fort Hood Independent Review Committee,* U.S. Army, November 6, 2020, p. ii).

tate the constant rotation of units into and out of both Operation Enduring Freedom and Operation Iraqi Freedom.[4] As operations in Iraq and Afghanistan wound down, the Army transitioned to the Sustainable Readiness Model (SRM).[5] SRM was similar to ARFORGEN, in that it represented a cyclic readiness model, where units built readiness over time, then had a period of employment, and then went through turnover that reduced readiness and placed them back at the beginning of the cycle. The Army's new model is the Regionally Aligned Readiness and Modernization Model (ReARMM). As the name implies, ReARMM is meant to improve on SRM by including specific time frames dedicated to modernization of units and by providing units a regional focus to help guide their manning, equipping, and training. In the words of one Army general:

> Army units operate in an environment of unpredictability, and arguably even instability. Units are placed on rotational missions based on their availability, and these missions vary in location, length, manning, readiness requirements and equipment just to name a few. Modernization today occurs when we can find a window to fit it in, or simultaneous with other activities. Every week, month and year is filled with constant change and high tempo for soldiers. Our soldiers and families can deal with a lot of tempo, but unpredictability results in an incredible amount of stress on the force.[6]

In some ways, the People First initiative serves as the counterbalance to the metrics-heavy force-generating models of the past that seemingly prioritized readiness numbers over the humans expected to achieve them. The challenge for the Army is that, in some ways, the imperatives of the People

---

[4] U.S. Army Forces Command (FORSCOM), "Army Force Generation," *STAND-TO!*, July 19, 2010; Charles C. Campbell, "ARFORGEN: Maturing the Model, Refining the Process," *Army Magazine*, June 2009.

[5] See Army Regulation 525-29, *Force Generation—Sustainable Readiness*, Headquarters, Department of the Army, October 1, 2019; and U.S. Army War College, *How the Army Runs: A Senior Leader Handbook, 2019–2020*, 2020, pp. 3-36–3-39.

[6] LTG Leopoldo Quintas, quoted in Andrew Feickert, *The Army's Regionally Aligned Readiness and Modernization Model*, Congressional Research Service, IF11670, Version 3, September 22, 2022a.

First initiatives and programs are at odds with the successful achievement of units' ReARMM goals and milestones. It is this friction between People First and ReARMM that we examine in this report.

Commanders of Army tactical units, at all levels from company level on up, must make decisions daily about how to balance their mission requirements with the need to provide soldiers with an appropriate work-life balance and professional development opportunities.[7] Our research revealed that, for each primary focus area in ReARMM, the decision criteria may vary. During a modernization phase, when units are turning in old equipment, receiving new equipment, and training on that new equipment, there may be a premium on access to supply and maintenance soldiers. During a training phase, there may be a premium on having cohesive teams training together through a lengthy sequence of progressively more difficult events. And, during a mission phase, whether deployed or merely maintaining readiness in a prepare to deploy order (PTDO) status, there might be a bias against allowing soldiers time off to be with family or pursue non-Army interests.

The Army's personnel system also impacts this process. The Army's emerging talent management personnel process is focused on longer-term readiness issues—fully developing the capabilities of soldiers by effectively matching their knowledge, skills, attributes, and desires with job assignments that are compatible.[8] If talent management is done well, the Army believes it will lead to more-effective soldiers and to better retention rates.[9] But this additional level of autonomy in job selection is not always compat-

---

[7] We note that one of our reviewers found that when one Googles "people first, mission always," the top results include the Army, the Air Force, the Navy, IBM, NASA, Harvard Business School, a bank, a manufacturing company, and Chick-fil-A. The balance problem that we discuss in this paper is not unique—although the nature of military life and Army life (in our case) adds some unique components.

[8] For information about the Army's talent management program, see both the Army's People Strategy (U.S. Army, *The Army People Strategy*, October 2019b) and the Army's talent management website and accompanying literature (U.S. Army Talent Management, homepage, undated-a).

[9] See, for example, U.S. Army, *U.S. Army Talent Management Strategy: Force 2025 and Beyond—Ready, Professional, Diverse, and Integrated*, Headquarters, Department of the Army, September 20, 2016, p. 8.

ible with Army needs to rotate people for a variety of reasons (even if the soldier does not want to move) or with the fact that not all soldiers can get their preferred job or duty location. Thus, the talent management process impacts which soldiers arrive in which units and when, and it is constrained by or responsive to factors that fall outside the realm of ReARMM. In short, the Army's assignment process cannot be as responsive as some units may require or desire. Although the system strives for stability in unit manning, the operational reality is that the imperatives (from multiple sources) for soldiers to rotate assignments can significantly affect the ability to provide that stability. The effect of this is that it is not uncommon for units to be missing key personnel at different points in their unit's ReARMM cycle or to have a limited number of those personnel available—impacting how commanders think about the work-life balance of those specific soldiers.

Another key Army goal is the development and maintenance of cohesive teams. In this usage, the Army is focused on more than stabilizing a crew, squad, or unit and allowing the same soldiers to work together and build teamwork. The goal of cohesive teams, in the words of General James McConville (Chief of Staff of the Army [CSA] at the time of writing), is as follows:

> There are three areas that I am most concerned about that are breaking trust with the American people and hurting our Soldiers: sexual misconduct, suicides, and racism. These are what building cohesive teams and the Sergeant Major of the Army's "This is My Squad" initiatives get after. . . . It's about having tough conversations to make sure that leaders understand everyone's perspective. More importantly, it's about getting to know each other's story and have a deep understanding and appreciation for each other.[10]

In this context, "cohesive teams" refers to the variety of leadership traits, as well as training requirements and culture change goals that the Army is pursuing. Meeting the goals of building cohesive teams requires a time commitment from Army leaders and soldiers that falls outside of the normal mission- or job-focused training events.

---

[10] James C. McConville, "People First: Insights from the Army's Chief of Staff," *Army Sustainment*, Vol. 53, No. 1, 2021, p. 21.

In short, commanders must balance soldier needs for work-life balance, personal and professional development, and time with family, as well as time for leaders to build cohesive teams with their soldiers, against the units' needs for soldier time and effort focused on unit mission training. It has also been well documented that the requirements placed on Army units far exceed the time available to achieve them.[11] Processes like ReARMM are designed to help give commanders and units predictability about what they will be doing and provide them guidance on where and when to take risks when not all tasks can be accomplished to the desired standard.

The key research question that FORSCOM asked RAND to explore was as follows: Are there ways to further reduce the challenges associated with managing this balancing act between mission (broadly defined) and work-life balance, especially given the Army's senior leadership focus on putting people first?

Methodology

Our research relied on six distinct levels or methods of research. We did policy-level reviews and interviews with both institutional organizations and troop units. We took a subset of the policy documents and the interviews and subjected them to a lexical analysis, using a tool called Dedoose. We used the troop interviews as the basis for development of a workshop that allowed us to examine some of the typical scenarios a unit might have with regards to People First and ReARMM frictions. Separately, we gathered Army personnel data and matched them to units of assignment and ReARMM cycle calendars to understand how ReARMM and a variety of personnel issues may be related. Finally, we looked, briefly, at academic and business literature to see whether there were useful insights from expectation management that the Army might be able to apply. These levels inter-

---

[11] See, for example, Lisa Saum-Manning, Tracy C. Krueger, Matthew W. Lewis, Erin N. Leidy, Tetsuhiro Yamada, Rick Eden, Andrew Lewis, Ada L. Cotto, Ryan Haberman, Robert Dion, Jr., Stacy L. Moore, Michael Shurkin, and Michael Lerario, *Reducing the Time Burdens of Army Company Leaders*, RAND Corporation, RR-2979-A, 2019; and Leonard Wong and Stephen J. Gerras, *Lying to Ourselves: Dishonesty in the Army Profession*, U.S. Army War College Strategic Studies Institute, February 2015.

acted in different ways to help us build out our understanding of the Army's challenges and to develop possible remedies.

Policy Review

Our policy review encompassed both formal Army documents (Army regulations, policies, doctrinal publications, etc.) and informal Army products—especially speeches and articles by the Secretary of the Army and the CSA, as well as Army websites for People First, talent management, and other related topics. Our focus was on policy and products related not just to People First or ReARMM but also things related to the Army's manning processes and to the policies for fielding new Army equipment (the focus of the modernization phase of ReARMM). The scope of the policy review can be seen in our bibliography. The policy review was critical in helping develop our interview protocols, and it continued to expand throughout the project as we identified new areas of inquiry or a need to better understand an existing one.

Interviews

We sought to capture a variety of perspectives regarding the risks to achieving the Army's ReARMM and People First goals. With the help of our FORSCOM sponsor, we coordinated interviews with 12 different organizations during the summer of 2022, including FORSCOM offices responsible for implementation of manning guidance, modernization and fielding, and ReARMM. For unit-level interviews, we talked with staffs and key leaders from FORSCOM units at the corps, division, and brigade levels that could be impacted by these programs. Our units represented a variety of functional and multifunctional brigades and both heavy and light combat forces. Finally, we also interviewed personnel at Headquarters, Department of the Army (HQDA), and other stakeholders within the Department of the Army from the personnel, training, and modernization communities. We conducted interviews via phone or video teleconferencing, all within a group setting. We developed a semistructured interview protocol that was adequately flexible to shift the substance of the conversation depending on the type of organization represented during the interview, though the overarching focus remained generally the same and covered the following aims:

- Gauge participants' understanding of the ReARMM and People First concepts.
- Understand participants' and associated organizations' role in implementing those concepts.
- Illuminate challenges and opportunities in implementing ReARMM and People First requirements.
- Elicit views on potential implications and mitigation strategies.

Lexical Analysis

We sought to identify themes and insights and their potential implications for REARMM and People First execution. We organized themes that emerged during the interview discussion applying the familiar doctrine, organization, training, materiel, leadership, policy, facilities, personnel (DOTMLPF-P) construct using Dedoose, a cloud-based qualitative analysis software program that facilitates team-based coding and subsequent data analysis.[12] We then categorized interviews by unit sizes and types to gain a sense of whether trends in perspectives differed depending on the echelon or type of organization represented in each discussion. We categorized unit type as command staff, functional, infantry, modernization organization, or personnel organization.[13] We categorized organizational type as either policy (HQDA organizations) or operational (division level and below organizations). We discuss our analysis and findings in Chapter 3.

Scenario-Based Workshop

We conducted a workshop to assess Army leader priorities of common Army tasks to assess mid-grade leaders' metacognition of People First implementation. We assembled a convenience sample of field-grade Army officers to participate in role-based scenarios to explore how these officers

---

[12] See SocioCultural Research Consultants, Dedoose, web application for managing, analyzing, and presenting qualitative and mixed-method research data, version 9.0.84, 2023.

[13] Modernization organizations are those organizations responsible for coordinating and executing the fielding of new equipment; personnel organizations are those organizations responsible for assigning (and reassigning) personnel.

worked through hypothetical friction points between People First initiatives and ReARMM requirements. For example, in one scenario, the role-players worked through an instance in which a soldier requested work schedule accommodations due to conflicts with unit regional training and the off-post child care center's day care hours. Such topics as spousal unemployment or underemployment and having enough time with family are emerging concerns impacting military families.[14] With this in mind, we designed our scenarios to include the soldier's spouse in half of our scenarios. The officers rotated playing such roles as a junior soldier, the spouse of the involved soldier, company and battalion commanders, and installation support services in a variety of types of combined arms battalions, medical functional bridges, and operational brigades. Prior to and after the role-based scenario workshop, the officers participated in a rank-order scaling exercise to understand potential impacts of the role-based workshops on how these officers viewed People First implementation. We analyzed the officers' pre- and post-workshop rank-order scaling using a frequency count of the instances in which officers increased or decreased their prioritization of time-off requests relative to other mission requirements. A detailed description of the workshop design and results is found in Appendix B.

Measuring and Predicting Friction Using Quantitative Analysis of Personnel and ReARMM Data

We approached the quantitative analysis in such a way that it would provide results that were most actionable in terms of addressing conflicts between People First initiatives, the ReARMM cycle, and the Army's manning and equipping process. Prescriptive analytics[15] of the macro systems (i.e., analysis to inform optimal schedules or courses of action) were unlikely to be implementable because this would require massive changes, such as rescheduling ReARMM across the Army. Moreover, our interviews with

---

[14] Blue Star Families, *2021 Military Family Lifestyle Survey Comprehensive Report*, 2022, p. 10.

[15] For definitions and descriptions of types of analytics, see Institute for Operations Research and the Management Sciences, "Operations Research and Analytics," webpage, undated.

personnel suggested that the conflicts were not at the macro level but were instead specific micro-level issues that are hard to predict or schedule.

On the other hand, applying descriptive analytics (i.e., what do the data tell us is happening?) and predictive analytics (i.e., can we expect certain friction points based on strong statistical associations?) appeared to be useful, because descriptive analysis would highlight the degree of problems (or the lack thereof) and provide areas of improvement for the Army to focus on. In addition, the predictive analytics might also advise commanders of when and the degree to which friction points will emerge over the ReARMM cycle to better prepare for them. Specifically, we performed statistical analysis to measure whether certain issues or friction points (e.g., military occupational specialty [MOS] mismatch, turnover in units, deployment considerations) were more likely to manifest in certain phases of ReARMM than others. In addition, we attempted to predict the effect (or lack thereof) of these issues on fill rates, retention, and other metrics of concern.

The quantitative analyses can serve to confirm or moderate friction points identified in the qualitative analysis of the interviews, while describing what friction points may be worse than others. Moreover, these analyses can suggest where impacts are the largest, quantify otherwise anecdotal observations, and illuminate potentially hidden challenges or causal links.

Expectation Management

As we explored various areas of friction across organizations and unit types, it became clear that part of the root cause of the issues we identified came down to the need to ensure that Army leaders and soldiers are applying a healthy dose of expectation management. To bolster our thinking on how the Army can mitigate this challenge, we conducted a noncomprehensive survey of potential mitigation strategies in the literature on this topic across different industries. We defined *expectation management* as a process for consistently communicating with key stakeholders (employees, bosses, clients, customer base) while shaping perceptions of one's intent, intended process, outcomes, and duration. We applied this literature to the Army culture and context to identify whether there were easily translatable lessons. In the context of the U.S. Army and ReARMM modeling, this translates into reducing friction not by eliminating the conflicts that might arise in scheduling or personnel availability, but rather by mitigating the morale effects of those conflicts as they occur.

Organization of the Report

In Chapter 2, we provide additional details on ReARMM, People First, and FORSCOM's and our understanding of friction and recap the research questions that our investigation of these three things suggested. Chapter 3 provides the details of our qualitative research. Specifically, it discusses our policy reviews, interviews, lexical analysis, scenario-based workshop, and examination of the expectation management literature. The chapter concludes with a discussion of our findings and recommendations from this aspect of the research. Chapter 4 complements Chapter 3 through quantitative analysis. We present our goals, data sources, and analytic structure and process. We present our findings, a discussion of their limitations, and the recommendations we derived from the quantitative results. Chapter 5 recaps our project and offers our detailed findings and recommendations, as well as suggestions for follow-on work. Appendix A contains the detailed regression modeling results and coefficients from our quantitative work done in Chapter 4. Appendix B provides further details on the scenario-based workshop and the literature on expectation management.

CHAPTER 2

Background on ReARMM and People First

Introduction

The primary focus of our research was on the friction occurring at the intersection of the Army's ReARMM process and its People First initiatives. This chapter presents background on how and why the Army came to adopt both processes, as well as some descriptions about the way each works (or is intended to).

ReARMM

The Emergence of ReARMM

Over the past 40 years, the Army has gone through three readiness models.[1] From the 1980s to 2006, the Army utilized the Tiered Readiness Model, in which some units were fully manned and equipped, while other, identically designed units received less than their full authorization for soldiers and equipment. Units therefore also trained at different levels of responsiveness, with the more fully resourced units expected to fight earlier in a contingency and the less fully resourced ones arriving later. This process worked well during the Cold War, with known adversaries (i.e., the Soviet Union

---

[1]   Andrew Feickert and Lawrence Kapp, *Army Active Component (AC)/Reserve Component (RC) Force Mix: Considerations and Options for Congress*, Congressional Research Service, R43808, 2014. We follow Feickert's definition of readiness: the military capacity to engage in combat and fulfill assigned mission and tasks.

and North Korea) and specified war plans. After the collapse of the Soviet Union, the system functioned—albeit with some strain—through the 1990s, as long-term operations in the Middle East and the Balkans required the Army to continuously rotate troops into and out of the mission areas.[2]

In 2006, ARFORGEN was adopted to man, equip, and prepare units for combat deployments in Iraq and Afghanistan. The focus of ARFORGEN was supporting any emerging needs for the rapid mobilization of forces to support the operational and global contingency demands of the geographic combatant commanders and civil authorities, while ensuring a steady supply of units—properly equipped and trained—to support the ongoing combat operations. Conceptually, for most Regular Army (RA) units, ARFORGEN consisted of a three-year cycle that included a *Reset* cycle, a *Train/Ready* cycle, and a *Deploy* cycle.[3] In Reset, soldiers returned from deployment to a "minimal manning and equipping profile" that allowed them to "take a knee." This period was marked by no significant requirements on the units to be "ready." As the cycle progressed, units gradually increased toward regular training standards. In Train/Ready, soldiers trained extensively to prepare for their next overseas mission, before transitioning to Deploy—where units were considered ready and available for deployment (and, in fact, most often did deploy).[4] An outline of ARFORGEN readiness requirements is provided in Figure 2.1.

In practice, however, the demand signals for Army forces created the need to accelerate rotations faster than the objective goals of once every three years for RA forces and once every five years for reserve component

---

[2] For an overview of the changing nature of Army readiness processes, see M. Wade Markel, Alexandra T. Evans, Miranda Priebe, Adam Givens, Jameson Karns, and Gian Gentile, *The Evolution of U.S. Military Policy from the Constitution to the Present, Volume IV: The Total Force Policy Era, 1970–2015*, RAND Corporation, RR-1995/4-A, 2020.

[3] Feickert, 2022a.

[4] For more on ARFORGEN, see Campbell, 2009; Alexandra Hemmerly-Brown, "ARFORGEN: Army's Deployment Cycle Aims for Predictability," *STAND-TO!*, November 19, 2009; and John M. McHugh and George W. Casey, *America's Army: The Strength of the Nation at a Strategic Crossroads—A Statement on the Posture of the United States Army, Fiscal Year 2012*, posture statement presented to the 112th Congress, 1st Session, U.S. Department of the Army, 2011, Addendum F.

FIGURE 2.1
Army Force Generation Model

The structured progression of readiness over time, to produce trained, ready, and cohesive units prepared for operational deployment in support of combatant commander and other Army requirements.

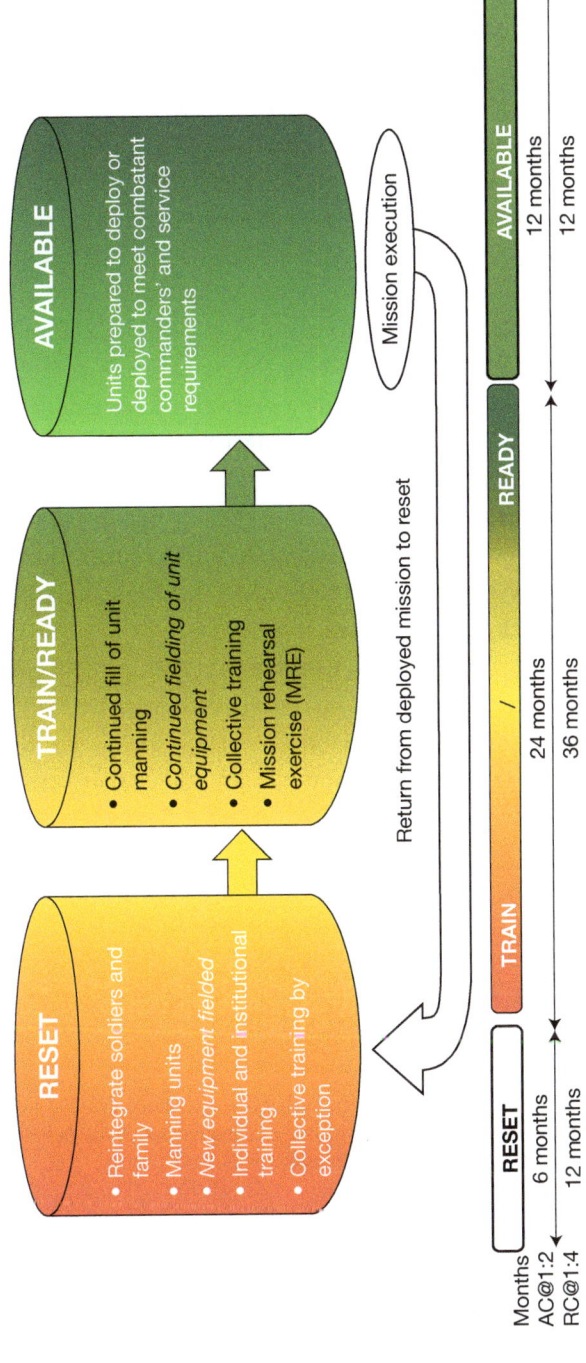

SOURCE: Adapted from Army Regulation 525-29, *Army Force Generation*, Headquarters, Department of the Army, March 14, 2011, Figures 1-1 and 1-2.

(RC) forces. Because of this, the length of the ARFORGEN phases, for many units, was shortened. They were generally still predictable, in the sense that units knew how long they would be in each phase and when and where to expect their next deployment.[5] ARFORGEN's supply-based model was effective for supporting an Army focused on fighting two ongoing wars. However, as the United States emerged from the two wars in Afghanistan and Iraq, and the security environment—and consequent demand for Army forces—changed in character, ARFORGEN was found lacking.[6]

By 2016, the strategic environment was beginning to shift away from large-scale deployments to two ongoing stability operations. Instead, the Army was starting to focus more on the potential need to engage in large-scale combat operations against major powers—specifically, concerns about Russian and China, combined with lingering concerns about North Korea. The Army believed that "the high and enduring operational tempo and limited expansion of the Army resulted in the degradation of the Army's readiness to rapidly respond to a large-scale wartime contingency with ready and responsive Army forces."[7] In response, the Army replaced ARFORGEN with SRM, which was implemented in 2017.[8]

Under SRM, the goal was to achieve two-thirds (66 percent) combat readiness of RA and National Guard brigade combat teams (BCTs) by 2023, with no fixed progressive cycles for active duty.[9] SRM's modules consisted of the *Mission* module, the *Ready* module, and the *Prepare* module, which, all together, made up a four-year process. This replacement was meant to provide the Army greater flexibility in prioritizing unit readiness and responding to contingency operations due to the lack of progressive cycles for active

---

[5]   For a good overview of ARFORGEN rotational policy and issues, see Timothy M. Bonds, Dave Baiocchi, and Laurie L. McDonald, *Army Deployments to OIF and OEF*, RAND Corporation, DB-587-A, 2010.

[6]   There were, throughout, several critiques of ARFORGEN. See, for example, Ricardo R. Garraton, *Analysis of Army Force Generation Model Behavior and Expectation Management*, U.S. Army War College, December 3, 2012.

[7]   Army Regulation 525-29, 2019, p. 3.

[8]   Mark A. Milley, "Memorandum for All Army Leaders, Subject: Army Readiness Guidance, Calendar Year 2016–7," Chief of Staff, Army, January 20, 2016.

[9]   Feickert, 2022a.

service units. In Mission, units are assigned a mission and are "validated, fully resourced, and immediately ready to conduct decisive action operations if required."[10] In Ready, units achieve or sustain a baseline level of decisive action proficiency and the ability to respond to contingencies. Lastly, the Prepare module consists of units reestablishing readiness while not committed to any mission.[11] This was meant to stabilize manning and avoid abrupt readiness declines, thereby sustaining longer and higher periods of readiness over time. In essence, SRM was designed so that unit readiness bands were narrower than those experienced under ARFORGEN. Units in Prepare were more ready than had been units in Reset under ARFORGEN. Units in Ready were more ready than had been units in Train/Ready, and units in Missionwere just as ready—but for both contingency and planned missions, as had been units in Deploy.

However, the lack of a fixed cycle impacted units by pushing units to "breaking" levels of readiness. In the words of one soldier, the Army's "relentless push" would result in wearing down the force, which would lead to some unit leaders inaccurately reporting readiness levels.[12] Reccurring issues included facing "challenges in staffing the [Army's] evolving force structure, repairing and modernizing its equipment, and training its forces for potential large-scale conflicts."[13]

The Army found itself experiencing an increase in instability and unpredictability. Irregular training cycles and rotational missions (location, length, manning, readiness requirements and equipment), along with an unpredictable modernization schedule, all led to soldiers experiencing a high operational tempo (OPTEMPO) and burnout. The Army's response

---

[10] Andrew Feickert, *Defense Primer: Army Multi-Domain Operations (MDO)*, Congressional Research Service, IF11409, November 21, 2022b.

[11] Alec Bannister, "Bravo Battery, 3-4 ADA BN 108th ADA BDE," *Spartan Magazine*, undated, p. 4.

[12] Haley Britzky, "Soldiers Say the Army's Relentless Push for Readiness Is 'Breaking the Force' in Leaked Documents," *Task & Purpose*, September 20, 2019.

[13] U.S. Government Accountability Office, *Army Readiness: Progress and Challenges in Rebuilding Personnel, Equipping, and Training*, testimony before the Subcommittee on Readiness and Management Support, Committee on Armed Services, U.S. Senate, statement of John H. Pendleton, director, Defense Capabilities and Management, GAO-19-367T, February 6, 2019.

was to update the SRM process and convert to the current ReARMM program.

ReARMM Defined

Following the publication of the National Defense Strategy and National Military Strategy in 2018, ReARMM was formally adopted in 2021 by the Army as a means to maintain global competitiveness by aligning forces and equipment regionally to support readiness requirements and prepare the force for the future.[14] Its implementation was meant to smoothly transition units through the three phases that make up ReARMM's life-cycle process, all while providing a regional alignment to help focus the mission training.

ReARMM is expected to better balance OPTEMPO and to correct past readiness issues resulting from significant demands placed on units, leaders, soldiers, and families. ReARMM consists of eight-month phases for *Modernization*, *Training*, and *Mission*. The Modernization phase focuses on necessary unit reorganization, fielding of new equipment, and training on that equipment. Next, in the Training phase, the units shift to training on unit mission-essential tasks and soldier training (e.g., weapons qualification, MOS training) necessary to support those unit-level tasks.[15] This is done in the context of an associated regional alignment. Training phases are followed by Mission, in which units are either sent on operational missions, are held in a PTDO status for a specified set of potential contingencies, or are just expected to sustain their readiness to deploy as a hedge against operational uncertainty and emerging missions. Ideally, these phases ensure a seamless transition of units between and across phases—having the right people on hand as they enter each phase. For most units, ReARMM will be a 24-month, three-phase cycle. However, some units will have four phases (a PTDO phase followed by

---

[14] Devon Suits, "Army Implementing ReARMM Unit Life Cycle Model," Army News Service, March 2, 2021.

[15] For purposes of our analysis and discussion, we only look at RA units. RC units go through the same phasing, but the phases have different lengths. We also note that ReARMM is still in a period between initial operating capability (IOC) and full operating capability (FOC). The Army expects that some time will be required to regularize all units into standard eight-month cycles. In the interim, many units have experienced cycles of different length or have been programmed to have them.

a deployment) or five phases that mix the basic three phases in different ways. Some units are identified as being on continuous employment (e.g., division headquarters) and are managed differently. Figure 2.2 provides a visualization of the goals of each phase.

Additionally, there are three key ReARMM principles that the Army seeks to improve with the new model: predictability, stability, and synchronization. The goal of predictability is to establish a stable rotation for units, soldiers, and family; the goal of stability is to create habitual relations with command and mission; and the goal of synchronization of the total force is as a means to increase predictability to better equip and integrate its force employment and modernization.

The new model was adopted because the Army suffered from readiness challenges caused by a number of factors involving personnel, equipment, training, a high OPTEMPO, and multiple other issues. Collectively, there was a perception that these factors resulted in stress for both units and the

FIGURE 2.2

Regionally Aligned Readiness and Modernization Model Unit Life Cycle

Re-ARMM unit life cycle model

Modernization (8 months)	Train (8 months)	Mission (8 months)
• Unit reorganization • Integrating modern capabilities • Displace equipment process • New equipment fielding/training	• Mission-tailored • Regionally focused • Individual/small unit/ collective training • HST/LFX/CTC/WFX	• Designated units assigned against specific missions/regions • On mission or ready for assignment • IRF/CRF/decision action

➤ Enables Army <u>transformational change</u> to a multidomain land power
➤ Focuses units regionally with **predictable, habitual relationships** to specific missions and theaters, enhancing Army support <u>**competition**</u>
➤ Synchronizes all Army components providing predictability to formations

SOURCE: Adapted from Kurt J. Ryan and Jin H. Pak, "Operationalizing ReARMM: A Sustainment Perspective," *Army Sustainment*, August 11, 2021.
NOTE: HST = home station training, LFX = live-fire exercise, CTC = combat training center, WFX = warfighter exercise.

individual soldiers. Box 2.1 describes the goals and benefits of ReARMM as articulated by the HQDA G-3/5/7.

This report was being researched and written at a time when ReARMM was in transition from its initiation (what the Army calls IOC) to its full implementation (what the Army calls FOC). As will be seen in later chapters, this means that policy is still evolving and that not all units have fully experienced a standard ReARMM cycle yet. In many ways, this report is part of the effort to continue to improve ReARMM as it moves to FOC and beyond.

People First

If readiness programs like ARFORGEN, SRM, and ReARMM lend themselves to consistent, detailed descriptions, the same cannot be said for People First. One theme constant throughout our study of People First is that it is many different things to many different people. We do our best to capture the broad range of ideas that have been discussed under the rubric of "People First" and perhaps, in so doing, also illustrate the roots of some of the challenges that Army leaders have in trying to understand exactly what activities should be prioritized or how to prioritize them in accordance with People First guidance.

In June 2021, the Secretary of the Army and the CSA made a joint statement before Congress announcing a shift in Army priorities and establishing a "deliberate achievable path to deliver a ready and modernized Army by 2028 and a transformed multi-domain Army by 2035."[16] In this testimony, they said:

> The Army's number one priority is now people. Our people are our Soldiers from the Active, Guard, and Reserve components, Army families, Army civilians, and retiree and veteran Soldiers for Life. At every

---

[16] Christine E. Wormuth and James P. McConville, "Statement by the Honorable Christine E. Wormuth, Secretary of the Army, and General James P. McConville, Chief of Staff United States Army, Before the Committee on Armed Services, United States Senate, First Session, 117th Congress, on the Posture of the United States Army," U.S. Senate, June 15, 2021.

BOX 2.1

ReARMM: Advantages and Continued Development

Advantages of ReARMM:

- Aligns Army units against regional priorities.
- Optimizes time available to plan, train and modernize.
- Creates predicable windows to field the modern capabilities necessary to build a multi-domain capable Army.
- Enables the Army to transform into a multi-domain force while providing a predictable supply of ready units—both for the Army and the Joint Force.
- Builds predictability for the Reserve Components, equippers, and personnel managers.

What Continued Efforts Does the Army Have Planned?

- ReARMM will facilitate consistent, manageable OPTEMPO, increased predictability for training and force employment, and prioritization of modernization.
- ReARMM aligns the Total Army against current Fiscal Year 2021/2022 (FY21/22) Joint Staff and Army requirements.
- At least three units of the same size, type, and modernization level aligned against a known Joint Staff requirement fulfill a "mission line."
- Dynamic force employment–style rotations (no overlap between rotating units) are the optimal solution for Competition requirements enabling the Army to manage OPTEMPO and allow Combatant Commanders maintain a strategically predictable, operationally unpredictable Army stance.
- Formations aligned under a mission line will have the same assigned modernization level to ensure common proficiency on fielded equipment.
- ReARMM does not create additional forward stationing actions or immediate changes to existing regional alignment.

SOURCE: Reproduced from Headquarters, Department of the Army, Deputy Chief of Staff, G-3-5-7, "Regionally Aligned Readiness and Modernization Model," webpage, October 16, 2020.

echelon, the Army must promote and build cohesive teams (1) that are highly trained, disciplined, and fit, (2) that are ready to fight and win, and (3) in which each person is treated with dignity and respect. Cohesive teams are the foundation of all our people initiatives and how the Army can best sustain readiness and transform for the future. Three critical enablers from the 2019 Army People Strategy continue to set conditions for putting people first: Army Culture, Quality of Life initiatives, and a 21st Century Talent Management System.[17]

The Army People Strategy referenced by the Secretary and Chief "describes how we will shift from simply distributing personnel to more deliberately managing the talents that our soldiers and civilians possess" and is depicted in Figure 2.3.[18]

If a single event could be seen as the driver of the Army's decision to prioritize people, it was the brutal murder of Specialist Vanessa Guillén at the hands of a fellow soldier in April 2020 at Fort Hood. The People First Task Force was founded in December of that year to implement the recommendations of the independent review committee that examined the incident.[19] The work of the task force to develop a plan to address these recommendations is ongoing.

The Army's public-facing website for People First specifically focuses on a set of initiatives to combat "issues and harmful behaviors that tear at the fabric of our force, including sexual assault, sexual harassment, suicide, discrimination and extremism."[20] In addition to the ongoing work of the

---

[17] Wormuth and McConville, 2021.

[18] U.S. Army, 2019b; and U.S. Army, "Army People Strategy: Overview," webpage, 2019a.

[19] The Fort Hood Independent Review Committee released a 136-page report in November 2020 that reviewed the command climate of units at Fort Hood, with a specific focus on sexual assault prevention and response, sexual harassment, and equal opportunity. For more information, see Swecker et al., 2020.

[20] U.S. Army, "Army People First: Prioritizing Our Most Valuable Asset—People First Task Force," webpage, undated.

FIGURE 2.3

The Army's Strategic People Approach Through 2028

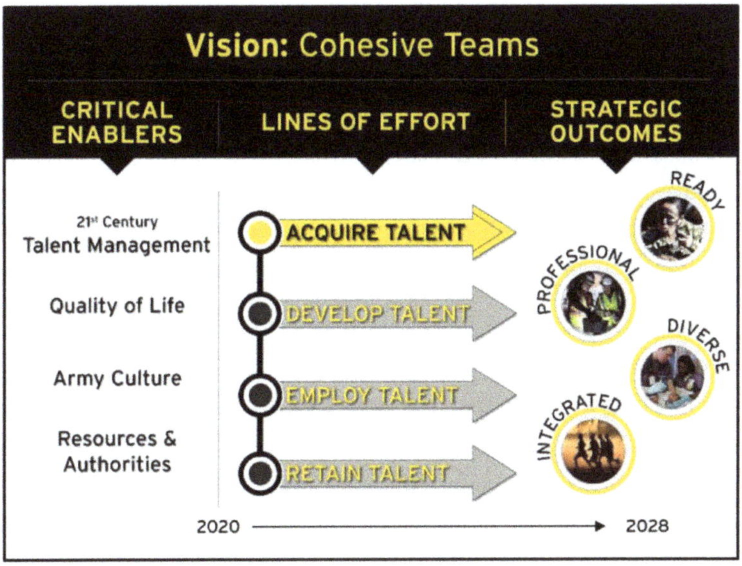

SOURCE: Reproduced from U.S. Army, 2019b.

People First Task Force,[21] the list of initiatives can be grouped into the following categories:[22]

- preventing sexual assault and harassment, through a planned redesigned of the Sexual Harassment/Assault Response and Prevention (SHARP) program

---

[21] The People First Task Force has implemented several Fort Hood Independent Review Committee recommendations and has made additional policy changes in response to the committee's work. For more information, see Michael Reinsch, "People First Task Force Building More Cohesive Teams," Army News Service, December 13, 2021; U.S. Army, "Secretary of the Army Announces Missing Soldier Policy, Forms People First Task Force to Implement Fort Hood Independent Review Committee (FHIRC) Recommendations," webpage, December 8, 2020; and U.S. Army Public Affairs, "Army Announces CID Restructure and SHARP Policy Improvements," webpage, May 6, 2021.

[22] U.S. Army, undated.

- suicide prevention, through the Army Suicide Prevention Program
- ending discrimination and extremism
- diversity, equity, and inclusion, through Project Inclusion
- junior leader development, through the This is My Squad initiative and the related Squad Leader Development Course
- talent management, through a planned redesign of the Army's talent management system and related initiatives
- quality of life, through a series of initiatives and policy changes across five categories: housing, health care, child care, spouse employment, and permanent change of station (PCS) moves.[23]

Although the Army's People First initiatives are distinct from the Army People Strategy, these initiatives conform to the broad categories of enablers defined in the document. The first five (along with the People First Task Force) are drivers of changes to Army culture. Fundamentally, Army leadership views culture as the combination of values, beliefs, and behaviors that drives the Army's social environment and that should continually change to align with the Army's strategy. The Army People Strategy describes the process of changing Army culture to meet a desired vision of cohesive teams as having three steps, consisting of:

1. building on positive aspects of Army culture while incorporating new cultural elements to meet the challenges of the Information Age (Define)

---

[23] PCS is when a soldier moves from one geographic assignment to another; expiration of time in service (ETS) is the date on which the soldier's enlistment contract ends and they are scheduled to be discharged from the Army. Other reasons for leaving could include retirement or being transferred within the same location to a different unit. While local commanders may have some flexibility on timing of on-post transfers, they have relatively little flexibility on timing of ETS or retirement and can only influence timing of PCS to a small degree. For both processes (moving or leaving the Army), soldiers have a set of entitlements beginning up to six months prior, which may require or allow the soldier to attend special programs or classes or provide time to take care of the movement logistics (e.g., selling their house or packing up their household goods). This takes some of the soldier's time away from the unit and represents demands that must be balanced with unit training demands.

2. relying on unit leaders to clearly define, communicate, and model the desired change in culture (Drive)

3. continual examination to ensure that the development of Army culture aligns with the *Army Strategy* (Align).[24]

A series of changes to the Army's talent management system seeks to develop the "policies, programs, and processes that recognize and capitalize the unique knowledge, skills, and behaviors possessed by every member of the Army team" through a data-driven, people-centric approach modeled after (and motivated by) the increasingly competitive domestic labor market.[25] Foundational among these is the Integrated Personnel and Pay System, a fully digital human resources system that seeks to provide greater transparency of soldiers' knowledge, skills, and behaviors; better manage personnel records; and integrate active, guard, and reserves into a single system. Other programs include the Army Talent Alignment Process, which creates a more market-style assignment process for officer assignments, and a variety of programs that provide assessments for rising mid-level officers and senior enlisted leaders that seek to determine their fitness for command.[26]

An ongoing series of policy and process changes, investments, and initiatives across the five categories of housing, health care, child care, spouse employment, and PCS moves makes up the Army's approach to addressing quality of life, a third key enabler of the Army's People Strategy. An ongoing overhaul of on-base residential housing and barracks renovation and replacement are central components of a plan to improve housing, while providing incentives to child care providers to address a critical need for Army families. Additional policies aimed at spouse employment, including reimbursement for professional licensing and certifications following a PCS

---

[24] U.S. Army, 2019b, p. 12.

[25] U.S. Army, 2019b, p. 2.

[26] The Army Talent Management Task Force is responsible for implementing a number of ongoing initiatives. For specific information, see U.S. Army Talent Management, undated-a.

move and earlier notification of PCS moves, aim to improve these aspects of soldiers' and their families' quality of life.[27]

Conclusion

As can be seen from the emergence of both ReARMM and People First, the Army clearly recognizes the dual nature of its challenges. The Army must maintain a modern, ready force that can both meet current (growing) operational demands and also be prepared for rapid deployment of ready forces to, potentially, very large and challenging crises at home and abroad. However, given its size, maintaining that readiness, combined with changes in the expectations and shape of the modern labor pool, is in tension with the ability to recruit and retain the Army's workforce—while establishing and maintaining a healthy work environment and organizational culture.

This broad range of concepts embodied in People First creates opportunities for a lot of nuances. There are some areas (e.g., suicide prevention) where the policy may be very clear and precise and others (e.g., work-life balance) where it may be far less so. People First, as essentially a work culture–focused set of guidance, contrasts with ReARMM, which is very much a process- and outcome-oriented program. The challenge is to find how well the two can accommodate each other. With this additional background, we now turn to the findings of our research into this tension.

---

[27] Latashia Bates, *Army Readiness and Modernization in 2022*, Association of the United States Army, Land Warfare Papers, No. 146, 2022.

CHAPTER 3

Analysis of Army Perspectives and Relevant Literature

Qualitative analysis provided us an opportunity to understand how soldiers and leaders experience ReARMM through a lens of People First and to better understand what they meant when they said that they were experiencing friction. In this chapter, we explain who we interviewed and what we learned from those interviews, as well as how a policy review helped us prepare for the interviews. We also provide our lexical analysis of both the policy review and the interviews. We then discuss how the interviews led us to conduct a role-playing workshop to help us better understand Army leader behaviors as they relate to the friction between ReARMM and People First. Finally, we present the findings that we synthesized from the full body of our qualitative work. In providing those findings, we organize them using the Army's DOTMLPF-P construct.

Interviews with Army Process Subject-Matter Experts and with Unit Leaders and Staffs

We sought to capture a variety of perspectives regarding the risks to achieving the Army's REARMM and People First goals, as laid out in the Army guidance documents discussed in Chapters 1 and 2. With the help of our FORSCOM sponsor, we coordinated interviews with 12 different organizations during the summer of 2022, including FORSCOM offices responsible for implementation of these programs, FORSCOM units at the division and brigade levels that could be impacted by these programs, and other stakeholders within the Department of the Army.[1] In preparation for the interviews, we also reviewed Army policy and briefings related to manning the forces, modernizing the force, and training the force.

We conducted interviews via phone or video teleconferencing, all within a group setting. We developed a semistructured interview protocol that was adequately flexible to shift the substance of the conversation depending on the type of organization represented during the interview. That said, the overarching focus remained generally the same and covered the following aims:

- Gauge participants' understanding of the ReARMM and People First concepts.
- Understand participant and associated organizational roles in implementing these concepts.
- Illuminate challenges and opportunities involved in implementing ReARMM and People First requirements.
- Elicit views on potential implications and mitigation strategies.

Lexical Analysis of Army Policy Documents and Interviews with Army Process Owners and Unit Leaders and Staffs

We coded and analyzed the interview data to identify themes and insights and their potential implications for REARMM and People First execution organized by the DOTMLPF-P construct. We used Dedoose, a cloud-based qualitative analysis software program that facilitates team-based coding and subsequent data analysis.[2] We then categorized interviews by unit sizes and unit types to gain a sense of whether trends in perspectives differed depending on the echelon and/or type of organization represented in each discussion. We categorized unit type using labels for command staff

---

[1] It is important to note that, due to time and resource constraints, we were not able to engage with soldiers below the brigade level, which would have provided insights into firsthand experience and perceived understanding of any ReARMM and People First dilemmas in implementation.

[2] SocioCultural Research Consultants, 2023.

(for division-sized units or smaller), functional (for garrison staffs), infantry, modernization organization (those with responsibility for ReARMM implementation, modernization, or readiness), and personnel organization (responsible for management and execution of manpower and personnel plans). Interviews were conducted with one command staff, two functional units, two infantry units and garrison staff, five modernization organizations, and two personnel organizations. In terms of unit level, we categorized units as either policy (HQDA organizations) or operational (organizations at the division level and below). In total, five interviews were conducted with operational units and seven with policy organizations. We then developed a codebook designed with three aims in mind:[3]

- to identify potential friction points associated with ReARMM and People First implementation across the DOTMLPF-P construct
- to gain an understanding of how policymakers and Army soldiers define ReARMM and People First initiatives and whether their definitions align
- to identify potential recommendations across the DOTMLPF-P construct for either ReARMM or People First initiatives.

The qualitative analysis team identified 210 excerpts across the 12 sets of interviews. Each excerpt could be tagged with multiple codes if an interview subject covered multiple topics in a single response, for a total of 1,063 distinct codes across the 210 excerpts.[4] We describe the substantive results throughout this chapter.

---

[3] One team member experienced in qualitative data analysis coded the interviews. We established intercoder reliability (that is, consistency and consensus in the application of codes) by having two other members of the qualitative analysis team code a selection of interviews and compared results with those of the initial coder. The coding team met weekly over the course of five weeks during this analysis to resolve any discrepancies in the application of codes.

[4] We recognize that command hierarchies in military organizations can lead to cultural norms in which a senior leader dominates the conversation. As such, we were mindful to consider when the conversation—and subsequently the coded excerpts— might have been skewed by a single dominant interviewee among a larger interview group (for example, in an interview with four subjects when one subject spoke more than 50 percent of the time). We identified three such instances in which a single inter-

Role-Playing Workshop with Field-Grade Officers

We also assembled a convenience sample of five field-grade Army officers to participate in role-based scenarios and informal surveys to explore how mid-level officers might work through a set of hypothetical friction points between People First initiatives and ReARMM requirements. Workshop participants served in combat support and combat service support positions, had both garrison and deployed experience, and collectively served in all three of the Army's components: RA, National Guard, and Army Reserves. We asked participants to explore four common but difficult work-life balance dilemmas. We also conducted informal surveys with participants before and after the workshop to first gain a baseline understanding of how mid-grade officers perceived the two initiatives before our engagement and, after the workshop, to ascertain their views after the concentrated effort to work through the intent and execution of the initiatives. The workshop provided interesting insight into how one cohort of field-grade officers prioritized People First initiatives when faced with a hypothetical conflict with ReARMM requirements.[5] The findings from these efforts are incorporated into the DOTMLPF-P implications described throughout this chapter. However, we note here that the workshop raised several interesting questions, the answers to which we think the Army may want to pursue:

- Would the changes in leaders' attitudes that occurred in a hypothetical scenario translate to real-life changes in leadership practices?
  - If so, how lasting would changes in leaders' attitudes be following training on People First initiatives?
- If leaders' positive attitudes toward People First initiatives diminished over time, at what frequency would maintenance training possibly be of benefit?

---

view subject could have skewed the findings of the interview coding effort. In these cases, we defined one respondent being responsible for more than double their share of coded excerpts relative to the number of individuals in the interview.

[5] The scenario-based workshop sampled a small subset of field-grade officers. The small sample size was not representative of the officer or Army Total Force population and was not sufficiently powered to analyze with robust statistical methods.

- If this kind of approach is of possible benefit, what method of delivery would best for maintenance training of People First initiatives?

Further details on the surveys and descriptions of the four scenarios are provided in Appendix B.

Training, Personnel, and Policy Findings

Examining across the DOTMLPF-P paradigm, our analysis found that interview discussions primarily centered on training, personnel, and policy implications associated with executing ReARMM and People First initiatives, followed by implications for leadership and materiel.[6] Organizations and facilities were mentioned less frequently, and Army doctrine not at all.[7] Figures 3.1 and 3.2 organize themes mentioned by unit type and organizational type, respectively, and show that functional units were most interested in discussing personnel and training implications. Modernization-centric organizations focused most on implications for training and policy. Command organizations somewhat evenly split the discussions between training, personnel, and policy. Infantry unit conversations were also roughly split among personnel, leadership, and training; this was also true for personnel-type organizations. These data suggested to us that Army leaders across echelons and functions have a somewhat consistent assumption of where most points will most likely reside across the DOTMLPF-P landscape.

Figures 3.1 and 3.2 tell us where interviewees wanted to concentrate the conversation; the emphasis fell on training, personnel, and policy concerns associated with the dual implementation of ReARMM and People First ini-

---

[6] Because the workshop took place toward the end of our research phase, time constraints precluded us from analyzing the discussion in our Dedoose effort.

[7] This omission may be in connection with the acceptance of ReARMM (the doctrine) as the framework inside of which the other comments were being made. We do note that the overlap between policy and doctrine is highly subjective. For example, training management, which is frequently discussed, could easily be coded as doctrine or training instead of policy.

FIGURE 3.1

DOTMLPF-P Implications of People First and ReARMM by Unit Type

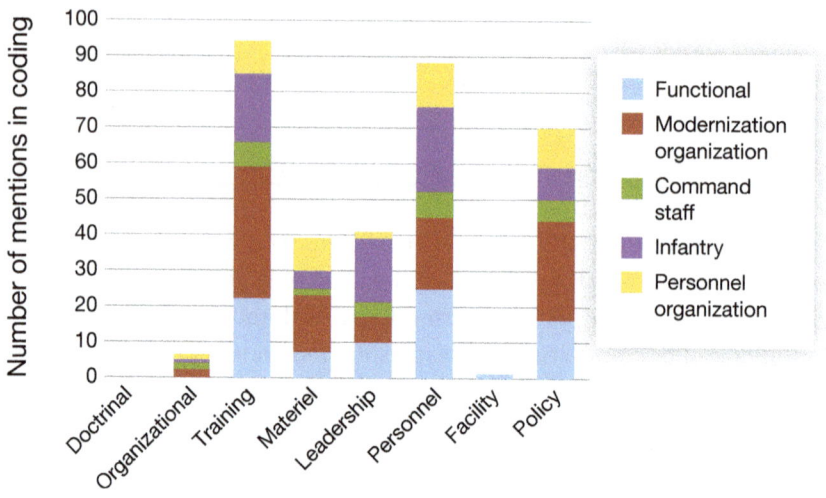

SOURCE: Authors' analysis using Dedoose.

tiatives, though materiel and leadership challenges emerged that will likely need to be addressed as well. The section below sheds light on the substance of these discussions.

Training Implications

Interview participants cited training-related challenges that they felt could occur as units carry out ReARMM and People First objectives. For example, modernization requires strictly sequenced timing to synchronize equipment with training schedules, but this leaves little flexibility to adapt when real-world events arise. In particular, comments highlighted how the war in Ukraine had disrupted unit activities, whether for training, modernization, or People First purposes. One infantry division staff member described the dilemma balancing requirements this way:

> Our mod [modernization] phase looks a lot like our training phase. We had another unit get pulled to fill a gap when [another unit] went to

FIGURE 3.2

DOTMLPF-P Implications of People First and ReARMM by Organization Type

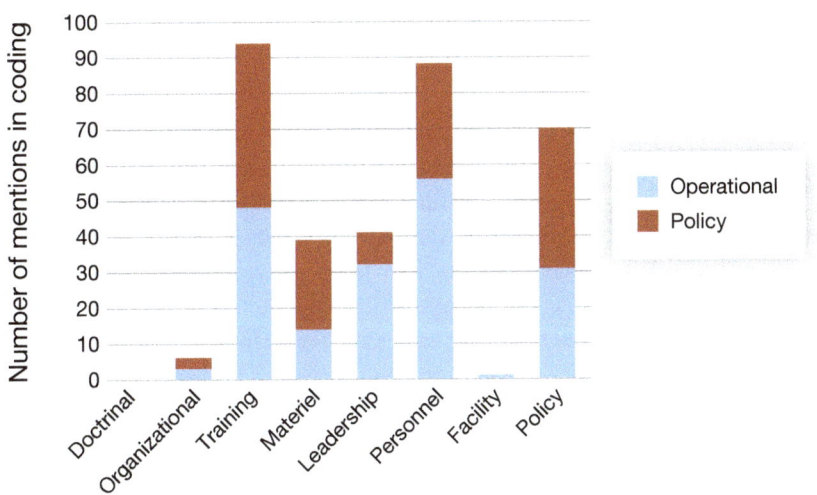

SOURCE: Authors' analysis using Dedoose.

Europe, which ruined their cycle. And that's just part of it, right? You want predictability, but you have real-world demands you have to fill. They're on a compressed cycle.[8]

Low-density units (Intel, Sustainment, Engineers, Medical) have their own unique challenges balancing training and ReARMM–People First requirements because of the constant deployment demands on a relatively small number of forces. One officer from an Intel brigade said:

As a brigade, we can't institute training cycles like our BCT brethren. . . . Since we've undergone a force design update, each of our battalions is in direct support role to division or corps HQ [headquarters]. Since a force design update, some of our units are designated ERF [European Response Force], so those aligned for ERF are on constant readiness so we can't take them out to train or modernize. And for People First,

---

8 Infantry division staff, interview with the authors, July 25, 2022.

if we're not giving them predictability on when they can stand down, we're not giving them predictability because we have to have consistent readiness.

A noncommissioned officer (NCO) from the same Intel brigade elaborated on the challenges:

> Within the past month, we were simultaneously shutting down operations in Europe [rotating out and placed on a PTDO to replace ourselves in supporting those capabilities to support us]. So we were doing both those things while going through NET/NEF [new equipment training/new equipment fielding][9] on [joint collaborative planning systems] and trying to do managed block leave. So training, fielding new equipment, conducting/closing ops and trying to rest.

Operational deployments are not the only dilemma to consider. Another infantry division staff member gave an example of how other unexpected training taskings can disrupt modernization requirements:

> There was cadet training this summer and that wasn't a planned event. One of our [brigades] had to assume that responsibility in like a month, and that responsibility takes a whole summer. That's just one example of the kind of stuff I'm talking about.

Additionally, incoming leaders have their own vision on training priorities that do not necessarily mesh with ReARMM time phases:

> These phases should be predictable, but it depends on when changes of command happen, other human factors that occur, right? A company or [battalion] commander comes in and they want their guys to train. . . . So they're coming in and not necessarily following the phases in the way they're supposed to.[10]

Our interviews suggested that ReARMM will run up against the inherently unpredictable nature of the Army's other roles, responsibilities, and

---

[9] NET/NEF is the Army's method for training and fielding new systems.

[10] Infantry division staff, interview with the authors, July 25, 2022.

requirements; leaders' discrete priorities might also detract from ReARMM goals. As one field-grade officer from our workshop noted, ReARMM may be "too brittle" to adapt to the needs of an agile Army.[11]

The trade-offs associated with meeting People First intentions while achieving training objectives were somewhat less clear to us. For example, according to our interviews with senior-level training officials, People First initiatives are less relevant for when units are training for or deployed to the combat training centers (CTCs). One branch chief from the training community explained that letting soldiers have time off with their families during a rotation (which is how the chief understood People First goals) simply would not happen because it would be "expensive and counterproductive" to the CTC's core mission to stress-test unit readiness. Additionally, providing time off during a rotation would elongate the actual time units would need to spend at the training center to ensure that they achieve the training objectives.

However, People First considerations can and do have an impact on CTC rotations, in indirect ways. For example, one training official described a recent situation where subordinate units of a division were experiencing fatigue after having been tasked with multiple deployments. Rather than pushing the rotation start date, the division commander requested sending only a limited portion of a brigade to an upcoming rotation. Army leadership "concurred, agreed and approved" the request.[12] This example demonstrates leadership committed to People First—taking care of soldiers' individual needs first, to enable improved unit readiness when needed later. We could not, however, discern whether the trade-offs associated with not having the full BCT complement at the training event outweighed the benefits of allowing the contingent of forces stay behind.

During our workshop, field-grade officers volunteered as role players in several notional scenarios that dealt with training dilemmas. In each scenario, they were asked to decide between effectively completing a training requirement or approving soldiers' individual requests to attend to personal pursuits. In scenario after scenario, discussions underscored some of the

---

[11] Comment from workshop participant, November 30, 2022.

[12] Army training official, interview with the authors, May 26, 2022.

difficult compromises both leaders and their soldiers must make to satisfy both personal and professional requirements (see Appendix B for details).

The tendency of role players was to first turn to fellow soldiers and family networks to help accommodate demands or to sacrifice personal and family needs for the needs of the unit (e.g., "Maybe an on-post neighbor can help with the child care gap?" "Might your anniversary date just push slightly to the left or right?" "Can we pull a trainer from another unit to substitute while our high-performing soldier attends career-enhancing PME [professional military education]?").[13]

Big Army was the last resort, and when participants did elevate dilemmas to fellow role players who were playing higher echelons of command, neither ReARMM nor People First initiatives seemed to adequately remedy the situation. ReARMM was no match for the inevitability of unanticipated events. As one officer contended, People First "briefs well on paper," but the pressures that officers experienced during the experiment seemed just as unsurmountable in the notional situation as they do in reality. As role players exhausted solution after solution, one participant ultimately expressed the collective frustration, "It is what it is."[14]

Materiel

Ideally, effective ReARMM implementation necessitates that new systems are planned, programmed, and fielded within the modernization window well in advance of a training rotation. However, interviewees raised important issues that will need to be worked out to ensure that the Army achieves its modernization objectives. For example, the challenges associated with matching the right people with the right systems at the right time will manifest in different ways across the force. We heard from members of high-OPTEMPO, low-density units that ReARMM poses unique dilemmas from

---

[13] Notably, and consistent with our Dedoose analysis of interview data, field-grade officers rarely mentioned doctrine, organizations, or facilities as relevant to the circumstances of a scenario. However, one officer suggested that garrison support (facilities) might be an option for covering a child care gap, noting the Army Community Services center.

[14] Army field-grade officer, workshop comment, November 30, 2022.

their vantage point. An NCO from an engineer brigade described the mis-match this way:

> The ReARMM cycle doesn't work for engineers. Our equipment align-ment doesn't align with the ReARMM cycle because of postponed equipment. . . . When they're [the units] already in their training or even mission cycle, they're getting their equipment, so they basically have to start their mod phase over. Our equipment fielding is all over the place, so that makes the standard ReARMM cycle very hard to achieve.[15]

Additionally, synchronization needs will not be the same for every unit, nor for the same reasons. For example, a senior garrison training staff member noted that while an infantry BCT's modified table of organization and equipment (MTOE) involves a substantial amount of equipment, infan-try units have far fewer equipment requirements and would likely not need an eight-month modernization period:

> I think the modernization phase is critical in heavy units. . . . In light infantry our modernization does not impact us as much. It's very small. You're going to get a new radio, OK. I mean, I'm trying to think of stuff we get, it's very small, it's down to the individual almost. So, I don't think having people at the right time is as big of a deal for us.[16]

However, another interviewee held the opposite position, warning against underplaying the challenges that are likely to come:

> These systems [may] be less complicated, but fielding a truck to a unit consumes a unit to get it done. You're moving trucks in and out, GFE [government-furnished equipment], NET/NEF. Etc. It's not specific to the major units, it's endemic across everything we field.[17]

---

[15] Functional brigade NCO, interview with the authors, July 28, 2022.

[16] Chief of training, infantry division garrison staff, interview with the authors, July 25, 2022.

[17] HQDA policy organization, interview with the authors, July 29, 2022.

Army personnel with whom we spoke had differing views on the impact that modernization will have on soldiers and unit readiness. This raises the possibility that matching manpower to materiel in a standardized modernization cycle may not be the most effective way to think about implementing modernization. That said, our interviews with a modernization-centric policy organization reminded us that ReARMM is still a work in progress, and, because the initiative "requires a level of synchronization that we haven't seen in the past,"[18] what "right looks like" has yet to be codified.

U.S. Army Training and Doctrine Command (TRADOC) is actively working through some of these issues. During our research, the command was in the process of developing a "scorecard system" to better match the number of personnel required in time for the delivery of a given NET/NEF package.[19] However, the policy official from the modernization organization acknowledged that even this effort may run into challenges because MTOE changes are sometimes announced late, which can make it difficult to ascertain the manpower requirements needed to field specific systems.

Personnel

A running theme in this report is that the Army may struggle to ensure that enough of the right people are available at the right time to meet ReARMM objectives. High-demand/low-density units stood out in our discussions as being particularly ill-equipped (or -manned) to meet expectations. For example, functional units noted that low-density MOSs such as sustainers (e.g., bridge crewmember, engineers) might struggle to provide the manpower required to meet ReARMM demands. One S1 officer questioned how feasible it is to commit to modernization expectations given current manpower constraints:

> How does ReARMM [factor in] the manning portion, do we have enough personnel to be able to support modernization? With our 12 Charlies, Novembers [engineer MOSs 12C and 12N], how does that

---

[18] HQDA policy organization, interview with the authors, July 29, 2022.

[19] HQDA policy organization, interview with the authors, July 29, 2022.

affect how we will receive the equipment if we don't have the right people there for NET/NEF?

An Intel officer raised similar doubts, noting constraints posed by continuous deployments:

> ReARMM doesn't apply to Intel. [We are] a small unit, no embedded redundancy to have any modernization cycles to focus on modernization, we are juggling everything all the time. Intel WfF [warfighting function] isn't for ReARMM. No deliberate cycling with modernization. We struggle all the time in the EMIB [Expeditionary Military Intelligence Brigade] [because we have] continuous employment.

With respect to soldiers' individual circumstances, during the workshop, a field-grade officer who played the role of a high-performing squad leader requesting time off for a wedding anniversary was quick to note that one of the benefits to ReARMM is that the routinization of phases will allow soldiers to know what is coming well in advance to avoid such conflicts in the first place. We note here that the squad leader and the individual playing the role were Army standouts who would likely already have the time management skills to avoid avoidable time conflicts. Yet, even the most competent soldier will run into challenges synchronizing ReARMM cycles with existing Army personnel programs, such as PME, PCS, or ETS. During the workshop, for example, role players struggled to find a happy compromise for a stellar supply clerk whose PME slot that he needed to attend for promotion happen to come up during his armored brigade combat team's (ABCT's) NET/NEF.

Real-world reflections were similar. A division staff member opined that the PCS "is too short to be flexible. If you miss a PCS cycle, you probably won't have a soldier from six months to a year."[20] Another staff member from a different division described similar constraints:

> Timing [of when soldiers arrive or leave] plays a big role. The only thing we have control over is commanders and SGMs [sergeants major]. We can do some adjustments to that. . . . But for others, if the unit's in the

---

[20] Infantry division staff, interview with the authors, July 28, 2022.

training cycle getting ready to go into mission cycle, if they [the new soldiers] come in at the end of the training cycle that doesn't do us much good.[21]

One command sergeant major from an infantry brigade suggested that Army Human Resources Command (HRC) needs to ensure situational awareness on how modernization cycles will impact solider training during key transition points:

> In a perfect world, transitions would happen in modernization. Then you are talking ARFORGEN. HRC should take that into consideration, where we are in the ReARMM cycle. That we would maybe help give some consideration for this. [The] easiest way [to manage the issue of soldiers transitioning out at critical moments of the unit life cycle] would be if HRC is aware of where we are and help us before we're on a training cycle. The reset into the training period is the real problem.[22]

We spoke with one G1 officer who acknowledged that modernization cycles are currently out of synch but expressed confidence that the process would eventually get there:

> Right now, it's all over the place, some are 8 months, some are 9, because it's at the very beginning. As the ReARMM model goes, the longer cycles are lessened down and will fit into 8-8-8. Depending on some equipment they're getting. 1-3 [BCT] was in modernization for 2 years because they're getting state of the art tanks, essentially a whole new fleet, so they are in modernization longer. . . . as ReARMM continues to move forward you'll start to see, I think starting FY24–25, everyone neatly fits into 8-8-8.[23]

The G1 officer's comment is a useful reminder that our research was taking place concurrently with the ramp-up of both ReARMM and People First and that some of the kinks in the system may be mitigated as

---

[21] Infantry division staff, interview with the authors, July 25, 2022.

[22] Infantry division and brigade staff, interview with the authors, July 25, 2022.

[23] Policy organization staff, interview with the authors, June 1, 2022.

FORSCOM stakeholders work through issues that surface. This was the intent, for example, of an upcoming (August 2022) Army People Synchronization Conference for relevant staff from Corps G1, Division G1, HRC, and TRADOC aimed at providing stakeholders an opportunity to cross-talk potential friction points. FORSCOM also hosts weekly meetings across the relevant stakeholder groups for a similar purpose. As one policy official explained, "We all come together and try to identify where those friction points are, come up with rules, various ways to make [the] system work not just for personnel, but for training, and all of that."

Leadership

Army leaders must set the tone for how subordinate units carry out guidance to execute these initiatives. However, the commitment to these initiatives by some seasoned leaders seems somewhat muted. With respect to People First, a commonly expressed sentiment was that taking care of soldiers is "just something good leaders do"[24] and that defined initiatives or policies were not necessary for good leaders to address their soldiers' needs. Indeed, we heard more than once that in many ways, ReARMM and People First initiatives are not that much different from what the Army has always done.

> I've been in 27 years and as far as training management is concerned nothing has changed really. Whether they're calling it ReARMM or something else, it's all the same, right? You come into work thinking you're going to do something and then something else happens, what you have to do changes, and you get on with it. It's really about the people that manage the system. There's all these fancy words people use to say the same thing we've been doing already forever. Leaders take care of their soldiers. Period.

One training chief at an infantry division explained that ReARMM and People First policies would not represent much of a change in the way the Army trains or develops its soldiers because the very nature of what the Army does requires soldiers to accept some level of uncertainty.

---

[24] Army field-grade officer, workshop comment, November 30, 2022.

> It's harder for the younger soldiers to realize this, that you have to be flexible. "Soldiers first," all these initiatives and stuff, it's for younger folks. If you've been in for 10 years, you get it. The roadmap here isn't going to be perfect, things change, stuff happens. But young kids, they're told, "You'll be home for Christmas." Older folks get it.[25]

These comments suggest that at least some mid-rank officers and senior enlisted leaders appear to downplay the notion that these two initiatives will have a notable impact on the force. Such perspectives can have important talent management implications. To this point, prior research has noted that overly tasked junior-level leaders already struggle to juggle readiness requirements to meet the priorities set by higher echelons.[26] Without consistent messaging from higher echelons on the measurable value placed on prioritizing People First principles, when soldiers request time for professional development pursuits that do not directly contribute to unit readiness, junior-level leaders may be less inclined to approve these requests.

Part of this reluctance is likely because People First principles are not factored into methods of evaluating a leader's performance that impact future promotions. One workshop participant noted the subjective nature of the way leaders measure the impact of People First goals:

> Some of the guidance is an idea. . . . Let your soldier go to his kid's baseball game. That's an idea and leadership style. There's no metric for how many ballgames you went to.[27]

One garrison staff official who works with soldiers transitioning out of the Army described the context this way:

> It's very competitive at the company level. These officers are rated at the battalion level. Some become myopic, they see the mission set as overwhelming rather than looking at the importance of success for individual soldiers. And you've got to realize there's no success measurement of "X number of my soldiers got their associate degree."

---

[25] Infantry division and brigade staff, interview with the authors, July 25, 2022.

[26] Saum-Manning et al., 2019; Zimmerman et al., 2019; Wong and Gerras, 2015.

[27] Army field-grade officer, workshop comment, November 30, 2022.

The interviewee asserted that failing to demonstrate commitment through clearly measurable goals serves as a detriment to both the soldier (whose requests are rejected) and the Army's retention goals (when that solider ultimately decides to leave the force):

> Soldiers say they're leaving since "the Army doesn't care about me," but of course that's not always true. I think we need to make changes, as difficult as it would be, and it would be [difficult] for me. But People First is just going to be a bumper sticker if you don't incentivize junior leadership, if you rate it differently.[28]

We recognized from our interviews and document reviews that People First concepts are inherently difficult to measure. Yet there are mechanisms that can be leveraged to useful effect. For example, several interviews suggested that the Army should identify opportunities to assess a leader's commitment to People First principles in officer efficiency reporting (OER) and noncommissioned officer efficiency reporting (NCOER) comments. As one garrison staff member contended:

> Having something that says, "List 8 People First accomplishments you oversaw" [in an OER] would go a long way.[29]

As we explored potential reasons why Army leaders and soldiers alike may encounter real constraints as they implement guidance, it became evident that some of the challenges might derive from differing interpretations of the policies and, as a result, variance in how the initiatives are applied. We discuss friction points as they relate to policy below.

Policy

We sought to establish a solid conceptual understanding of the People First and REARMM policies as described in Army policy and guidance. Of the two initiatives, we assessed that REARMM was the more straightforward.

---

[28] Division garrison staff, civilian employee, interview with the authors, July 25, 2022.

[29] Infantry division garrison staff, civilian employee, interview with the authors, July 25, 2022.

ReARMM guidance establishes clear mission priorities between modernization, training, and a variety of mission types that can be directly tied to Army doctrine. For example, Army Regulation 220-1, which lays out explicit guidance on the readiness status reporting metrics and requirements for Army organizations, installations, and units, offers Army leaders tangible concepts for assessing whether a unit has achieved milestones associated with the phases of modernization.[30] This type of process lends itself well to traditional quantitative or optimization analysis (as seen in Chapter 4).

People First concepts, however, as previously mentioned, are more challenging to measure, in part because personnel interpret the concept differently. For example, when we asked the staff of one functional brigade what People First meant to them, responses varied among participants in significant ways. One staff officer described his commander's interpretation of People First policies as providing soldiers predictable time off:

> 1500 release on Friday. Nonnegotiable. Very positive. Within 8 weeks the brigade level commander must approve anything [changing in the training schedule]. We've got it in our orders and doctrine.

In contrast, another staff member from the same brigade interpreted People First in the opposite direction and through the lens of readiness:

> A happy soldier isn't one who wants to take off at 1400. If I can do well, that's what I see as genuine happiness. Not mowing the grass, but doing the job I signed up for. . . . When troops are out and have the right equipment, they will compete on their metrics. Get their score good, for a training environment. They don't mind missing dinner if that's what they want to do.

Still another brigade officer focused on professional development—even when that education is not necessarily germane to the Army:

> [There is a] mutual understanding for people going to college and Army. They're not always pushing dirt, and bidding their time, and we

---

[30] Army Regulation 220-1, *Army Unit Status Reporting and Force Registration—Consolidated Policies*, Headquarters, Department of the Army, August 16, 2022.

help them, they are happy. Understanding them and their concerns, they will perform.

To understand why Army personnel derived such different meanings from the term "People First," we turned to Army doctrine and examined how various echelons of FORSCOM units incorporated their understanding of People First into guidance. We found that Army organizations defined People First themes differently and placed inconsistent degrees of priority on them. Table 3.1 demonstrates this trend through a single chain of command, from FORSCOM to XVIII Airborne Corps, to two of the Corps' subordinate units, the 82nd and 101st Airborne Divisions. Table 3.1 provides our extract of each commander's enumerated priorities for their command for FY 2022.

As Table 3.1 illustrates, despite FORSCOM placing the highest priority on People First concepts, subordinate units' policies do not reflect this emphasis. Definitions of People First–related concepts also varied among the units, seeming to reflect a lack of shared understanding on what encompasses People First. FORSCOM provides the most comprehensive definition by including the solider, civilians, family, and leadership development as a combined top priority. At the lowest end of the spectrum, the 82nd Airborne Division communicates a singular focus on "paratrooper first," though articulated as the fourth of four priorities of the command.

The varying definitions and priority given to People First initiatives in policy and guidance were an important finding in our research, suggesting that concepts can be interpreted and thus implemented differently in practice. This finding further amplified the idea that despite the focus People First might receive rhetorically, readiness continues to be the highest priority of lower-level leaders, as it is what they are incentivized to strive for and what is communicated to them by their leaders. **Put simply: The message of "People First" is not reaching the officers and staff NCOs in charge of managing and training soldiers with anything near the priority that the Army's senior leadership has set.**

The Army is not completely unaware of this challenge. Several interviewees described (or provided) a slide created by the CSA, GEN McConville. General McConville's thoughts are captured in Figure 3.3, which categorizes both work and professional responsibilities by his view of their level

TABLE 3.1

People First Emphasis in Selected Commanders' Guidance Documents

Unit	Priority 1	Priority 2	Priority 3	Priority 4
FORSCOM	People: Care for Soldiers, Civilians, and Families; Strengthen Leader Development	Readiness: Master the Fundamentals; Deliver Decisive Total Army Readiness	Modernization: Empower and Execute Reform; Inform and Implement the Future Force	
XVIII Airborne Corps	Readiness for large-scale combat operations	Lethality (at small unit level) Squads, Sections, and Platoons	Leader Development and Talent Management	Care of Soldiers, Civilians, and Families
101st Airborne Division	Contingency Resource Force Readiness	Establish People First at the Tactical Level	Innovation ISO Air Assault Operations	Leading with Fires
82nd Airborne Division	Masters of Joint Forceable Entry–Army	Transform the Division	Live the Airborne Culture	Paratrooper First

SOURCES: Features information from 101st Airborne Division Annual Training Guidance, FYs 2022–2023; 82nd Airborne Division Annual Training Guidance, FYs 2022–2023.
NOTE: ISO = in support of.

of importance relative to each other. For example, soldiers and their leaders should consider daily training as a routine event, similar in importance to carrying out personal appointments and school sports practices. On the opposite side of the spectrum, the chart suggests that a soldier being with family for occasions of life and death are on par with the ultimate duty and sacrifices of war.

While the opposing sides of the spectrum are somewhat easy to discern, our research suggests that subjectivity in leadership styles and approaches will make the middle section more difficult to decipher. Nevertheless, General McConville's attempt to "visualize the vision" sends a strong message to subordinate leaders about the need to help solidify the inherently fuzzy concept. This kind of rubric can help leaders and soldiers better understand how friction challenges should be resolved, which will also feed into consis-

FIGURE 3.3

Senior Leader Guidance on How to Conceptualize Work-Life Balance

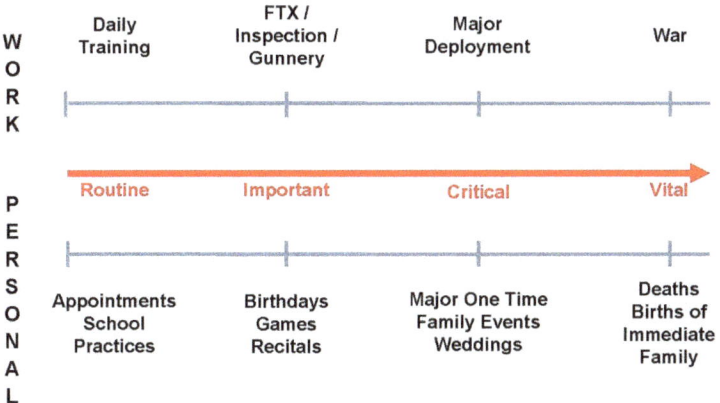

SOURCE: Reproduced from Haley Britzky, "Army Chief of Staff Gen. James McConville Wants Soldiers to Have Work-Life Balance," *Task & Purpose*, October 15, 2020.

tency in expectations. The Army needs to consider how to expand on this guidance.

One thing we did not find, however, was an effective feedback mechanism from units back to policymakers. On the one hand, the very existence of this research project suggests that FORSCOM (at least) is aware of the challenges, but it also suggests that the feedback loop may not be very responsive. An infantry division officer relayed how the lack of communication on how to effectively employ these concepts can frustrate the force:

> I think part of issue is at some level, whether it's necessary or not, it feels like you're being lied to, and I don't think that's done on purpose but it's how it might feel. I mean, you look at this piece of paper and it

tells you what you're going to do, but it's not what happens because we live in reality and things change. That could come off as dishonest.[31]

This indicates to us the importance of the Army developing feedback mechanisms that can routinely help leadership understand how ReARMM and People First are being understood and executed at unit level. Whether done through interviews or by more formal surveys or focus groups, the Army needs to "hear" from the soldiers more often and more granularly—and then communicate back to the soldiers how their concerns have led to change (or why they have not).

One promising observation during our workshop experiment suggests that when the People First message is reinforced, soldiers can and do shift their thinking. Prior to the workshop, participants were asked to fill out an informal survey on how they would prioritize a series of Army taskings and personal obligations. The cohort demonstrated a noteworthy shift in how they rank-ordered People First Initiatives immediately following the workshop, favoring increased priority to giving soldiers requested time off over mission requirements.[32]

Conclusion

Our qualitative analysis focused on understanding how soldiers and their leaders experience the friction between People First and ReARMM and how they are motivated to react to those experiences. Our key findings include the challenges of messaging and of incentives. In short, Army messaging is mixed, and, as one gets to lower-level units, messaging is increasingly more focused on near-term unit-level readiness or goals than it is on the overarching goals of People First. Even when the messaging gets through, the incentives are still aligned with training and the unit.

---

[31] Infantry division staff, interview with the authors, July 25, 2022.

[32] Our analysis of the workshop surveys showed six instances, out of seven scenarios, in which a participant gave greater priority to leave after the role-playing than they did in the pre-workshop survey. And in the post-workshop survey, there were only two instances in which a participant gave lesser priority to a time-off (leave) request than they had in the pre-workshop survey.

We note that external macro systems (the Army's personnel and modernization processes) are impactful, but we generally find that those systems have little incentive and few degrees of freedom to fundamentally change their processes. This means that the burden of dealing with the challenges remains on unit-level leadership. We also note, and many of our interviewees highlighted, that uncertainty and unpredictability are hallmarks of Army life—they are embedded in the nature of being a force responsive to outside stimuli (e.g., an unexpected need to send units to respond to a crisis in Ukraine). What this reinforces is that the communications challenge is not just one focused on clarity and consistency of messaging from Secretary of the Army all the way down to Bradley crewmember. Army leaders must be adept at, and practice, techniques of communication meant to help soldiers manage expectations and deal with that uncertainty. Doing so in the context of People First and its priorities will be a messaging and cultural challenge that the Army must confront.

In the next chapter, we offer some quantitative analysis that provides a glimpse of analytic techniques that the Army may be able to pursue to identify some places where systemic, rather than cultural, change may provide some opportunities to alleviate or mitigate friction.

CHAPTER 4

Quantitative Analysis of People First and ReARMM Friction Points

This chapter describes a set of quantitative analyses we performed to determine whether we could identify where the friction points between People First goals and elements of the ReARMM are most likely to manifest. We assess the amount of friction for various metrics of interest, such as the degree of conflicts across the stages and their impact on assignment fill rates, retention, and unit cohesion, among others. We begin by outlining the rationale for the types of data analytics we performed, as well as for some we chose not to consider. We then provide definitions and rationale for each dimension of analysis, followed by their results, and the degree to which those results point to statistically meaningful conclusions. Based on the results, we highlight specific findings and some key recommendations from the quantitative analysis, including augmenting or modifying our qualitative analysis of the previous chapter's interviews.

Goals of the Quantitative Analysis

We approached the quantitative analysis in such a way that it would provide measurable results and—depending on the specific result—likely indicate policy recommendations to reduce conflicts between People First initiatives and the ReARMM cycle. The macro systems that intersect with ReARMM (i.e., modernization and personnel) have limited ability to adapt to the ReARMM and People First friction points. This results in little room to deconflict scheduling in ways that reinforce People First. In addition, our interviews did not indicate that there was any evidence of widespread scheduling problems for requirements and resources. Even if the Army would consider major ReARMM cycle changes in response to People First concerns, these large, generalizable events appeared to be well managed from most perspectives. As a result, traditional prescriptive analytics of these systems (i.e., analysis to inform optimal schedules or courses of action) were unlikely to be either necessary or implementable.[1]

On the other hand, descriptive analytics (i.e., what do the data tell us is happening?) and predictive analytics (i.e., can we expect certain friction points based on strong statistical associations?) would be useful because they would highlight both the degree of problems (or the lack thereof), along with providing areas of improvement for the Army to focus on. In addition, the analytics might also advise commanders of when and the degree to which friction points will emerge over the ReARMM cycle to better prepare for them.

The quantitative analyses can serve to confirm or moderate friction points identified in the qualitative analysis of the interviews, while describing what friction points may be worse than others. Moreover, these analyses can suggest where impacts are the largest, quantify otherwise anecdotal observations, and illuminate potentially hidden challenges or causal links. Before describing the dimensions that we analyzed, we provide a list of the datasets we utilized for our analyses, how they were used to perform analysis, and their limitations.

Datasets Utilized and Their Limitations

Our quantitative analysis relied on four data sources: FORSCOM ReARMM Training Calendars, Total Army Personnel Database for Army Enlisted (TAPDB-AE), Army Training Requirements and Resources System (ATRRS), and FORSCOM MTOEs. Analysis was conducted on data dating from January 2020 to September 2022 (these dates were constrained by available ReARMM cycle data) and consisted of records for approximately 190,000 soldiers. Since the ReARMM cycles affect FORSCOM BCTs, the

---

[1] For definitions and descriptions of types of analytics, see Institute for Operations Research and the Management Sciences, undated.

quantitative analyses in this chapter are limited to those units. Table 4.1 summarizes the data excerpts and analytical utility of each database; further description of the databases follows.

ReARMM Training Calendars were provided by FORSCOM. These data are visual calendars that record the historical and anticipated ReARMM cycles from January 2020 through September 2022 for each FORSCOM BCT. These data were foundational to the analysis as they allowed us to map other databases to ReARMM cycles and explore personnel similarities and differences between the cycles. The goal of our analysis was to explore potential patterns among typical units; therefore, we excluded units that operate in specialized ways that are not representative of a typical FORSCOM unit—namely security force assistance brigades and multidomain task force units.

TAPDB-AE is an HRC database that includes service members' names, scrambled individual personnel numbers, and units. The data also include information related to service members' demographics, orders, promotions, assignments, training received, organizational affiliations, and readiness-related and deployment information, including assignment considerations

TABLE 4.1

Databases Used in Quantitative Analysis

Database	Filters	Utility
ReARMM calendars	1/2020–9/2022 35 FORSCOM brigades, excluding security force assistance brigades and multidomain task force units	Scopes the unit and range of analysis, identifies ReARMM training cycle dates for integration with other databases
TAPDB-AE	1/2020–9/2022 Individuals in ReARMM cycle brigades	Provides People First and readiness factors including grade, duty assignments, family, medical, standards, and conduct variables to link with ReARMM cycles; provides personnel levels to calculate fill rates (compared with MTOEs)
ATRRS	1/2020–9/2022 Individuals in ReARMM cycle brigades	Provides dates of education course attendance to link with ReARMM cycles
MTOEs	FY 2020–FY 2022 ReARMM cycle brigades	Provides authorized personnel levels to calculate fill rates (compared with TAPDB-AE)

and limited medical and behavioral/conduct information. This database provided our analysis with the People First and readiness variables to link with ReARMM cycles. It also allowed us to calculate unit fill rates by grade by comparing with MTOE data. TAPDB-AE data include enlisted data but do not include data on commissioned officer or warrant officer personnel.

ATRRS is an online Army Information Management System that manages and resources training courses attended by Army personnel. It includes training through multiple Army and Department of Defense programs. The ATRRS database covers training requirements, programs, costs, and personnel-level information to use in scheduling and filling classes to train Army soldiers. These data were linked with ReARMM data to explore patterns in education course attendance across ReARMM cycles.

FORSCOM MTOEs are based in current doctrine and policy and provide authorized levels of personnel by grade and position. MTOEs are updated each FY. Comparing assignments from TAPDB-AE to MTOE levels allowed us to calculate fill rates and explore patterns by ReARMM cycle.

In the next section, we describe in detail the seven dimensions of quantitative analyses we performed, all of which broadly fall under the definitions of predictive or descriptive analytics.

Definitions and Rationale for Each Dimension of the Quantitative Analysis

Our objective in the quantitative analysis was to measure and predict as many aspects of potential conflicts between the ReARMM cycle and the People First initiatives as possible with the data available and the time frame of the study, while also illustrating how these methods could be applied to additional analyses in the future. It was critical for each dimension of the analysis to cover one or more of three noted elements of People First: cohesive teams, talent management, and work-life balance.[2] While the dimensions we analyzed do not cover every aspect of these goals, they can serve as quantitative proxies for elements of the goals.

---

[2]   McConville, 2021.

In addition to informing elements of People First, most dimensions aimed to measure effects of the issues examined over ReARMM cycles: Modernization, Training, Mission–PTDO, and Mission–Deployment. While some of the analytical results for some dimensions do not necessarily provide actionable recommendations, they provide information on friction points, including where they may be more or less significant.

Dimension 1: Match Between Soldiers' MOS and Duty MOS over ReARMM

This dimension assessed within each FOSCOM unit the magnitude of the match between a soldier's primary, secondary, or additional MOS and the soldier's assigned duty MOS over the ReARMM cycle. This comparison also included Special Qualification Identifiers (SQIs) and Additional Skill Identifiers (ASIs). A high degree of *mismatch* would be a concern for cohesive teams[3] and talent management goals and ultimately could potentially affect retention, promotion, and readiness. In addition, understanding whether those mismatches occur more frequently during certain ReARMM phases (e.g., deployment) could help to inform the timing of readiness issues.

Dimension 2: Emergence of Specific Soldier Characteristics over ReARMM

We assessed the number of soldiers within a FORSCOM unit who became affected by special considerations (including assignment considerations and delay or deferral of assignment) related to medical considerations, failure to meet standards, conduct, and family and how those rates varied over the unit's ReARMM cycle. Understanding whether these considerations emerged more frequently during certain phases can help commanders pre-

---

[3] As discussed in Chapter 2, the Army's focus on cohesive teams involves combating toxic behaviors in the ranks. The Army's vision is to do this by ensuring that squad leaders and other junior leaders know their people and have opportunities to help them socially cohere—not just train together. In this sense, we view indicators of turbulence or of soldiers mismatched between their assigned jobs and the jobs they were trained for as challenges to this goal of cohesion.

pare and inform their decisions about specific cases. These issues are related to People First goals of cohesive teams and work-life balance.

Dimension 3: Differences Between a Unit's MTOE and the Number of Assigned and Available Soldiers over ReARMM

Building off the analysis in Dimension 2, if certain assignment considerations appeared more frequently in certain phases of ReARMM, we aimed to measure their impacts—if any—on personnel shortfalls during those phases. We analyzed the differences between the number of soldiers specified in the MTOE for a unit and the assigned and available soldiers over the ReARMM cycle to see whether differences were greater in certain phases, notably ones associated with higher degrees of soldiers with special assignment considerations. We performed analyses on the overall brigade fill ratio and broke down the fill ratios by grades: E4 and below, E5, E6, E7, E8, and E9. This dimension helps further address the goals related to talent management and cohesive teams and is directly related to readiness.

Dimension 4: Match Between Soldiers' Preferred and Actual Duty Locations

This dimension analyzed the degree to which a soldier's actual unit (and location) matches the soldier's stated preference (when given) and whether that varied significantly over the ReARMM cycle. Mismatches in general, or the degree to which mismatches were greater in certain ReARMM phases, may inform quality-of-life goals. In addition, mismatches, when assessed for association with other metrics, such as promotion and retention, may inform initiatives related to talent management.

Dimension 5: Conflicts Between Education Courses and ReARMM

We analyzed soldier attendance of education courses across each phase of the ReARMM cycle to see whether certain phases have greater levels of education attendance than others. This would be an indication of soldiers not being available to the unit in executing the focal purpose of the

ReARMM cycle (e.g., not able to help during modernization because they were in school). While the quantitative analysis might indicate differences, higher or lower levels of education attendance during a certain ReARMM phase may not point to a definitive set of reasons. For example, higher levels of education attendance during modernization could result from optimal timing and planning, or they could result from denied education courses during other phases, such as deployment. Understanding the differences— if any—over ReARMM helps address issues related to talent management, work-life balance, and cohesive teams. In this analysis, "education attendance" refers to a soldier attending any resident course between 15 and 180 days. We did not include courses less than 15 days in length or distance learning courses because of their lower disruption of ReARMM activities. Courses longer than 180 days are normally associated with a PCS, which we assess in Dimension 6.

Dimension 6: Conflicts Between Unit Turnover and ReARMM

This analysis measured the number of soldiers joining or leaving a unit (for example, due to an ETS or PCS) within a unit's ReARMM cycle to see whether there are measurable differences among phases. We analyzed the data for gains and losses of soldiers within each unit, resulting in various turnover metrics. While turnover is expected, unit cohesion and continuity may be more critical during certain phases. This analysis helps assess goals related to cohesive teams and talent management. In addition, unit readiness could be more impacted if turnover occurs more frequently during later, inopportune phases (e.g., late in the training phase, just prior to or during deployment).

Dimension 7: Associations Between Retention and People First Metrics

The analyses in many of the prior dimensions look at potential conflicts or issues over the ReARMM cycle (e.g., special considerations for family issues, matching between MOS and duty MOS). This final dimension extended the analysis of potential conflicts to examine the degree to which conflicts are associated with outcomes related to retention. While retention is based on a variety of factors,[4] we analyzed association, controlling for as many exogenous variables as possible, to provide insights into the People First proxy metrics (covering elements of cohesive teams, talent management, and work-life balance) analyzed in the prior dimensions.

Table 4.2 summarizes the breadth of coverage for the seven dimensions across the key People First goals. Again, these metrics and their corresponding analyses serve as proxies and are not intended to be a comprehensive representation—either singularly or as a group—for that goal. Instead, they represent what we believe are the best possible representations of those goals given the available data and the potential for new or additional insights that—to our knowledge—have not been analyzed previously, at least within the context of ReARMM.

TABLE 4.2

Dimensions of Quantitative Analysis Relation to People First Goals

Dimension	Cohesive Teams	Talent Management	Work-Life Balance
D1: MOS/duty MOS mismatch	X	X	
D2: Special considerations	X		X
D3: MTOE vs. assigned or available soldiers	X	X	
D4: Duty location preference match		X	X
D5: Education attendance conflicts	X	X	X
D6: Turnover conflicts	X	X	
D7: Associations with retention	X	X	X

---

[4] See Richard Buddin, *Success of First-Term Soldiers: The Effects of Recruiting Practices and Recruit Characteristics*, RAND Corporation, MG-262-A, 2005; and Bruce R. Orvis, Christopher E. Maerzluft, Sung-Bou Kim, Michael G. Shanley, and Heather Krull, *Prospective Outcome Assessment for Alternative Recruit Selection Policies*, RAND Corporation, RR-2267-A, 2018.

Results for Dimensions of Quantitative Analysis

This section provides an overview of the results for the seven dimensions of analysis defined in the prior section. While not all dimensions provide obvious policy takeaways, both the existence and the absence of statistically meaningful differences in certain outcomes across ReARMM can be illuminating and have implications for People First. As noted in the prior section, Dimension 7 is an extension of some of the People First considerations outside of the ReARMM cycle to other impacts of interest, such as relationships to retention. As a result, while some dimensions may have few takeaways in terms of their ReARMM impacts, we investigate some of them for broader impacts on metrics of concern to the Army.

The results in this section are presented at a high level. The details of the models employed, regression equations, and other identifying statistics are provided in Appendix A.

Dimension 1 Results: Match Between Soldiers' MOS and Duty MOS over ReARMM

Our regression analysis indicated that there was a high degree of MOS match (whether primary, secondary, or additional MOS) with a soldier's duty MOS across units. The level of MOS match was statistically significantly lower for deployment than it was for the modernization phase, albeit the decrease is modest: 96.8 percent for modernization and 96.1 percent for deployment. Figure 4.1 provides the level of matching across the phases at the aggregate level. We performed a t-test to assess whether the differences across the cycles were statistically significantly different from the "base" case of modernization (shown in white). We chose modernization as the baseline because issues that arise in this phase are generally less likely to be problematic than in later phases. As a result, whether the personnel issue or friction identified in that dimension decreases or increases relative to the modernization phase would be of potential interest to policymakers. We looked at MOS match from a high level across all MOS in the FORSCOM brigades; details are provided in Appendix A.

We denote a match at the MOS level to be a primary match. To investigate secondary- and tertiary-level matches, we add first SQI and then ASI to

FIGURE 4.1

Percentage of Soldiers Whose Primary, Secondary, or Additional MOS Matched Their Duty MOS, by ReARMM Phase

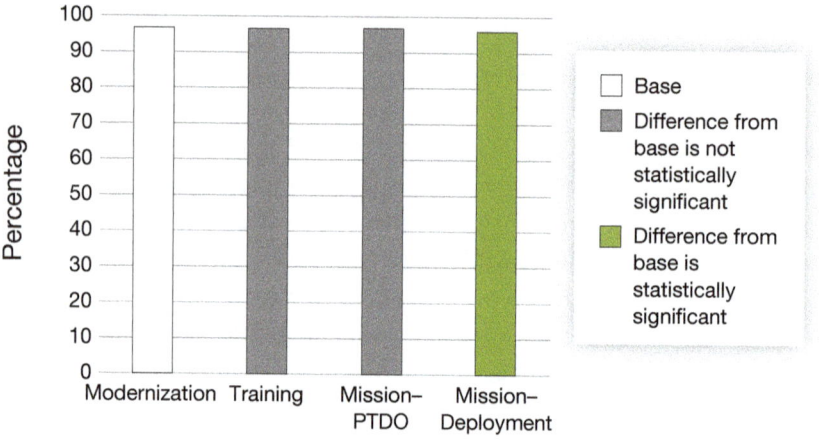

SOURCE: RAND Arroyo analysis of TAPDB-AE and ReARMM data.

the percentage of MOS matches. Naturally, the percentage of soldiers whose duty MOS *and* duty SQI (where duty SQI is specified) matched their primary, secondary, or additional MOS and SQI, respectively, decreases. As is the case with the MOS-only match, the only phase of ReARMM that was statistically significantly different from modernization is deployment, where the percentage matching both MOS and SQI was 91.3 percent, more than 1.5 percent lower than during modernization (92.9 percent). These results are shown in Figure 4.2.

When the analysis is further expanded to consider ASI matches (where a duty ASI is specified), mean match rate decreased to approximately 81.0 percent during modernization. Interestingly, the only statistically significant difference was an *increase* in the match rate during PTDO (82.3 percent) relative to modernization, as shown in Figure 4.3.

While beyond the scope of this study, understanding whether specific duty MOS or SQI were more prone to mismatches and whether certain primary MOS were used to fill gaps could potentially help leaders target specific improvements for certain specialties to create opportunities for HRC to manage those MOS differently.

FIGURE 4.2

Percentage of Soldiers Whose MOS and SQI Matched Their Duty MOS and SQI, by ReARMM Phase

SOURCE: RAND Arroyo analysis of TAPDB-AE and ReARMM data.

FIGURE 4.3

Percentage of Soldiers Whose MOS, SQI, and ASI Matched Their Duty MOS, SQI, and ASI, by ReARMM Phase

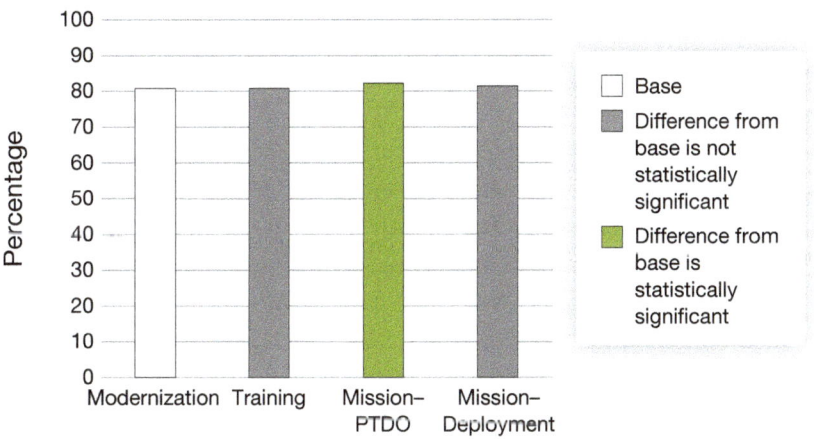

SOURCE: RAND Arroyo analysis of TAPDB-AE and ReARMM data.

Dimension 2 Results: Emergence of Soldier Assignment Considerations over ReARMM

We explored whether the number of soldiers affected by special assignment considerations varied significantly over the ReARMM phases. The special considerations could affect assignment considerations to include the delay or deferral of assignments. Gaps in units' assignment rates relative to the MTOE will be analyzed further in the next dimension of analysis.

We categorized certain assignment considerations from TAPDB-AE into four categories: medical, conduct, standards, and family. The specific variables that went into each category are provided in Appendix A, but, at a high level, we limited our choice of relevant variables to those that reflected issues related to *new* considerations rather than existing considerations before the soldier joined that unit. Figure 4.4 illustrates the percentage of soldiers within FORSCOM units with one or more assignment consideration variables for each category over the ReARMM cycle. Again, we performed a t-test to assess whether the differences across the cycles were statistically significantly different from the base case of modernization.

The results indicate that medical considerations were statistically significantly more common during PTDO than during the modernization phase. Conduct and family considerations were more common (again, statistically significantly) during training than during modernization. Standards considerations were lower during training, PTDO, and deployment than during modernization. In most cases, these differences across ReARMM—while statistically significant due to large numbers of observations—were not large in absolute terms.

Some of these results may not be surprising. During the more demanding training cycle, family issues that occur may be more of an issue than during modernization, while being resolved by the mission phases. Similarly, failure to meet standards during modernization might result in those soldiers being able to improve to meet those standards during the training or mission phases; otherwise, they may no longer be part of the unit during deployment for those reasons. As shown in Figure 4.5, for any phase of ReARMM, fewer than 10 percent of soldiers with one of the four assignment consideration areas had a second or third issue arise (no soldiers had all four consideration areas). These results may help commanders anticipate changes in personnel challenges and which ones are likely to manifest across ReARMM phases.

FIGURE 4.4

Percentage of Soldiers with Assignment Considerations, by ReARMM Phase

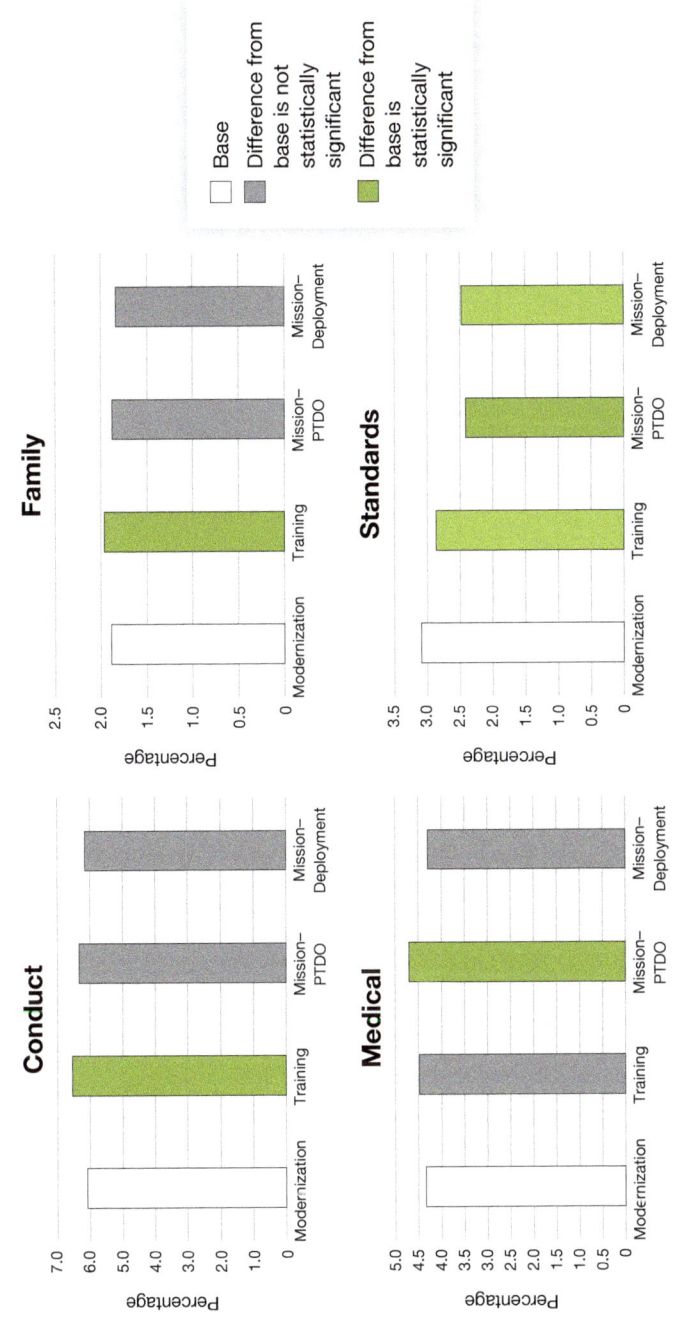

SOURCE: RAND Arroyo analysis of TAPDB-AE and ReARMM data.

FIGURE 4.5

Percentage of Soldiers with One of More Assignment Consideration Areas, by ReARMM Phase

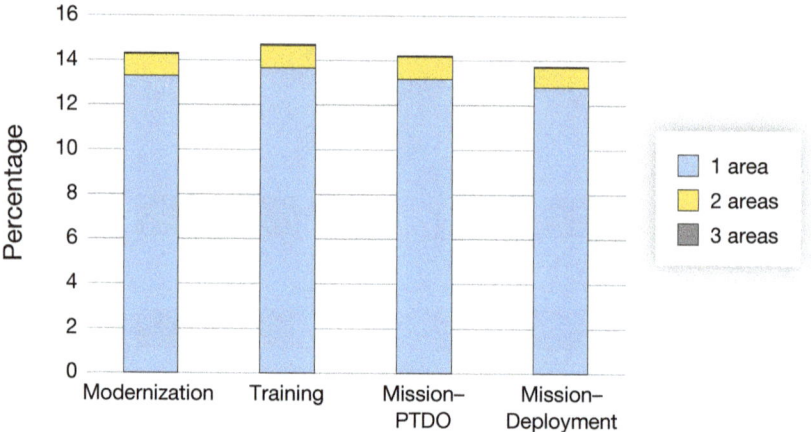

SOURCE: RAND Arroyo analysis of TAPDB-AE and ReARMM data.

Dimension 3 Results: Differences Between a Unit's MTOE and the Number of Assigned and Available Soldiers over ReARMM

This dimension examines the unit fill rates for enlisted soldiers across the ReARMM cycle. Based on the results of the Dimension 2 analysis, we wanted to see whether the higher (or lower) rates of assignment considerations manifested in different fill rates of soldiers as a percentage of those provided in the MTOE.

As Figure 4.6 illustrates, the overall fill rate during mission–PTDO was statistically significantly higher than during the modernization (base) phase, by about 2 percent. The other phases were not statistically significantly different than modernization. While special considerations for conduct and family issues are statistically higher in the training phase, they did not manifest in lower overall fill rates during that phase (perhaps offset by statistically lower rates of considerations for standards). This result would imply—at least at the aggregate level—that assignment considerations do not appear to meaningfully affect fill rates, except insomuch as their reduction during mission–PTDO for standards may be positively correlated with

FIGURE 4.6

Enlisted Position Fill Rates Relative to the MTOE for Units, by ReARMM Phase

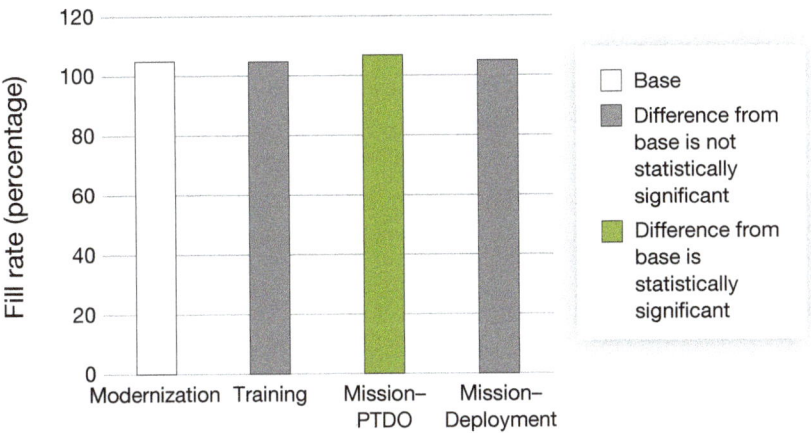

SOURCE: RAND Arroyo analysis of TAPDB-AE and ReARMM data.

a higher fill rate. It is important to note that we look at presence in the unit, not deployability ("available," as the Army terms it). So, while the conduct and family issues do not remove the soldier from the unit (i.e., presence does not change), the soldier may not be "available" (deployable) and able to participate fully in the mission of that unit. In all cases, the fill rate was higher than 100 percent of the MTOE, which is not necessarily surprising given the prioritization placed on BCT units. While beyond the scope of this study, it could be that other, non-BCT units did not display this relatively consistent pattern.

In addition to the overall fill rates, we analyzed the fill rates by enlisted grade, which reflect different patterns from the aggregate level. The details, along with the coefficients for the data presented in Figure 4.6, are provided in Appendix A.

Dimension 4 Results: Match Between Soldiers' Preferred and Actual Duty Location

This dimension analyzed differences between soldiers' stated location preference (either by state or by installation) and their actual duty location. In addition to understanding the level of preference match, we looked to see whether that level differed significantly over the ReARMM cycles.

The preference data records represent a relatively small percentage of the number of soldiers (just over 30 percent), but the number of data points is large (over 3.1 million data points) given the frequency of preference solicitation.[5] Figure 4.7 provides the level of preference matching for those soldiers expressing a preference across the ReARMM phases at the aggregate level for FORSCOM units. The match of preference to duty location was statistically significantly lower during mission–deployment compared with the modernization phase, while the other phases were not meaningfully dif-

FIGURE 4.7

Percentage Match Between Preferred and Actual Duty Location When Preference Was Provided, by ReARMM Phase

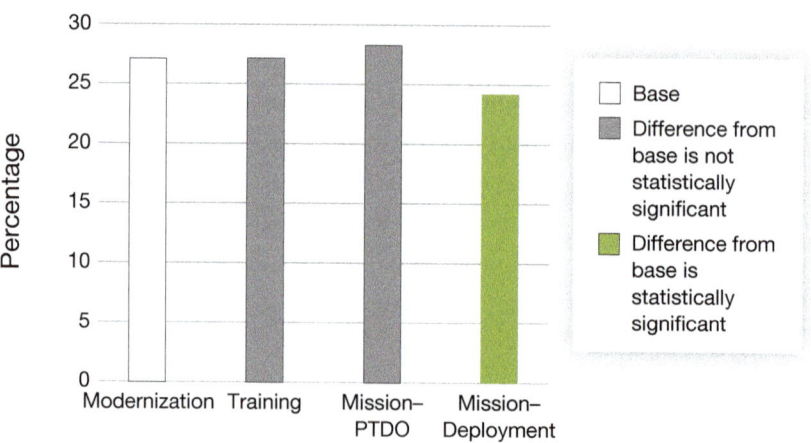

SOURCE: RAND Arroyo analysis of TAPDB-AE and ReARMM data.

---

[5] In other words, only 30 percent of soldiers in TAPDB-AE had an entry for preference. We do not know how representative that 30 percent is with respect to the soldier population we analyzed.

ferent from modernization. This result is not entirely unexpected, as it can be assumed that HRC will typically prioritize filling deploying units over meeting individual soldiers' location preferences. While the implications for ReARMM may be limited, this matching may be an important factor for retention, conduct, and performance indicators, in addition to providing commanders with insight into possible second-order effects of the relative location preference decrease of their soldiers during deployment.

Dimension 5 Results: Conflicts Between Education Courses and ReARMM

We analyzed soldier attendance of education courses across each phase of the ReARMM cycle to see whether certain phases have greater levels of education attendance than others. While education courses are part of the soldier experience, when that coursework occurs within the ReARMM cycle, it can affect unit cohesion. In this analysis, "education attendance" refers to a soldier attending any resident course between 15 and 180 days.[6] We break the analysis into two categories: courses that last between 15 and 89 days ("shorter courses"), and courses that last between 90 and 180 days ("longer courses"). The overall proportion of soldiers attending courses during the modernization phase was very small: around 3.3 percent for shorter courses and only around 0.5 percent for longer courses. The training phase saw a slight increase in these numbers for both shorter and longer courses, rising to 3.8 percent and 0.65 percent, respectively. During mission–deployment, the proportion of soldiers attending shorter courses dropped to around 2.4 percent. There was no significant difference in education attendance of longer courses during the mission–deployment and modernization phases. There also was no statistical difference between education attendance of both shorter and longer courses during the mission–PTDO and modernization phases. That there is a higher percentage of course attendance during training than modernization may be something the Army may wish to mitigate, though additional analysis on MOS-specific attendance would be

---

[6]   We do not include courses of less than 15 days or distance learning courses because of their lower disruption of ReARMM activities. Courses of more than 180 days are normally associated with a PCS, which we assess in Dimension 6.

needed to understand the relative disruption on each phase. These results are shown in Figure 4.8.

Dimension 6 Results: Conflicts Between Turnover and ReARMM

This dimension of analysis looks at turnover in units due to soldiers joining or leaving (e.g., PCS, ETS) within a phase of ReARMM. Turnover is inevitable, but some phases of ReARMM may be better able to handle that turnover than others. For example, turnover may be easier to deal with during the modernization phase than during a deployment. To explore this, we examined whether unit gains and losses are statistically significantly different between ReARMM phases.

Figures 4.9 and 4.10 show aggregate unit monthly gains and losses across ReARMM phases; the results reveal a few statistically significant differences. Compared with the modernization phase, units experienced both greater personnel gains and greater personnel losses during the training phase. During deployment, units experienced fewer personnel gains than during modernization. However, even when results are significant, the differences were small (differing by less than half of 1 percent), though this does not necessarily signify that unit stress points are absent.[7] For example, additional analysis could explore whether specific grades or MOSs are more susceptible to losses (or gains) than others, which could affect a unit's ability to perform certain tasks or missions. By extension, certain MOSs may be more critical during specific phases of ReARMM, and variations in their presence may be of interest.

The timing of unit gains and losses *within* a ReARMM phase may also be important. For example, turnover that occurred during the early portion of a training phase is likely less disruptive to unit cohesion and readiness than turnover that occurred toward the end of a training phase, when the ability to integrate new soldiers is reduced. Figures 4.11 and 4.12 plot monthly gain and loss rates across the fraction of the phase.[8] The figures illustrate

---

[7] See Appendix A for a full report of statistical results.

[8] For instance, if a unit's ReARMM phase lasts eight months, there will be data points for gain (or loss) rates at 0.125, 0.25, 0.375, 0.5, 0.625, 0.75, 0.875, and 1.0 for that unit.

FIGURE 4.8

Percentage of Soldiers Attending Education Courses, by ReARMM Phase

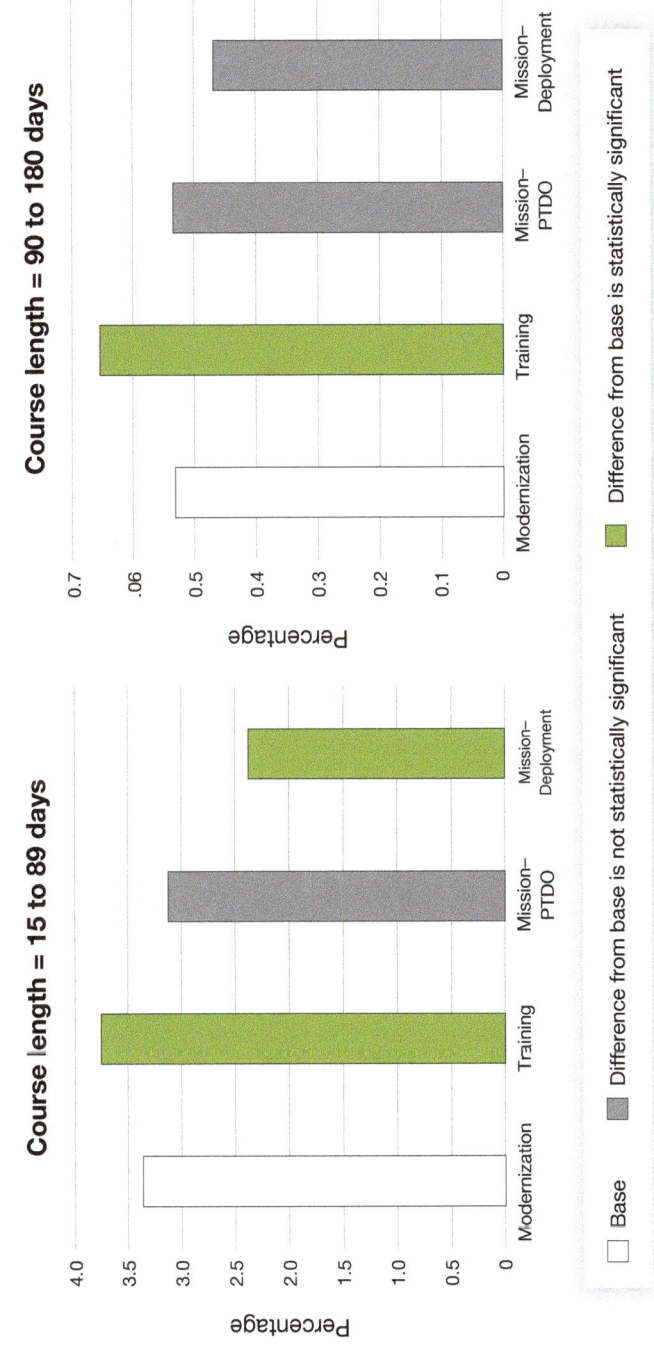

SOURCE: FAND Arroyo analysis of TAPDB-AE and ReARMM data.

FIGURE 4.9

Average Monthly Unit Gain Rate, by ReARMM Phase, as Percentage of Personnel

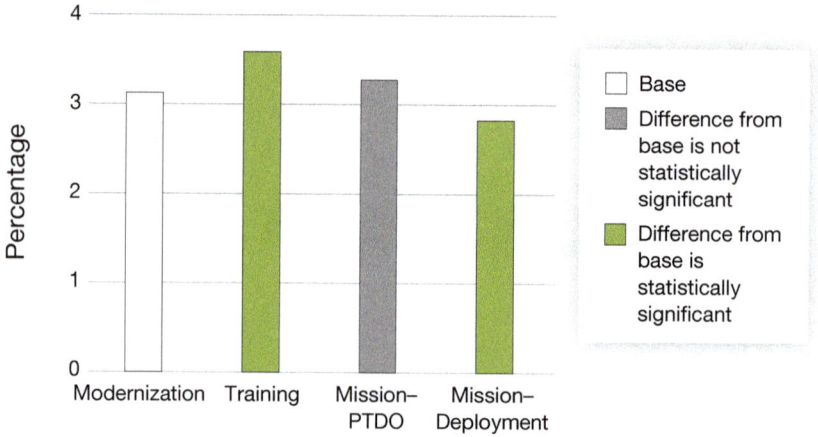

SOURCE: RAND Arroyo analysis of TAPDB-AE and ReARMM data.

FIGURE 4.10

Average Monthly Unit Loss Rate, by ReARMM Phase, as Percentage of Personnel

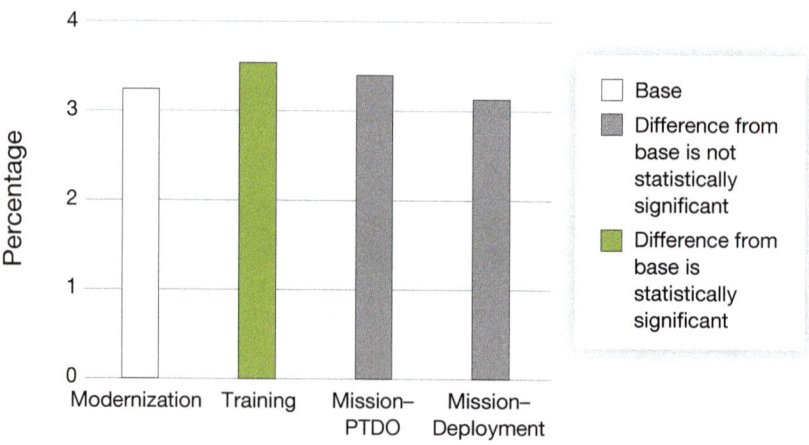

SOURCE: RAND Arroyo analysis of TAPDB-AE and ReARMM data.

FIGURE 4.11

Monthly Gain Rates for Units, by ReARMM Phase, as Percentage of Personnel

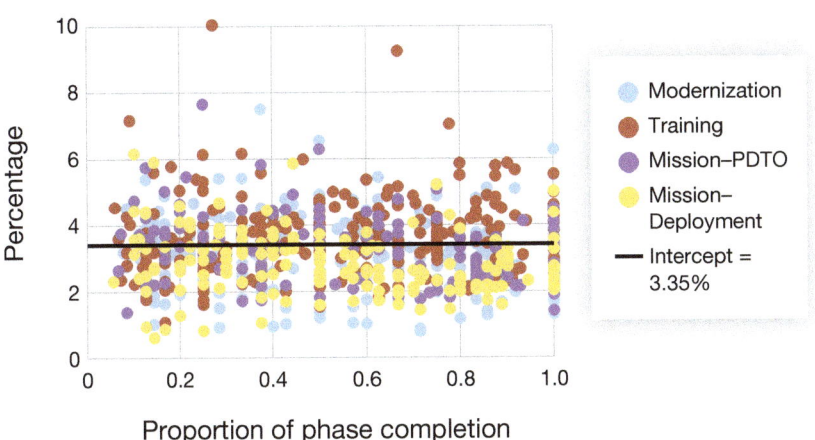

SOURCE: RAND Arroyo analysis of TAPDB-AE and ReARMM data.

FIGURE 4.12

Monthly Loss Rates for Units, by ReARMM Phase, as Percentage of Personnel

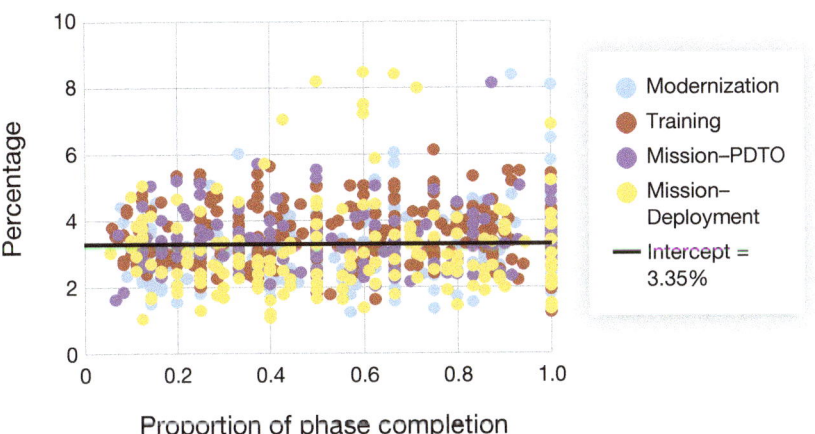

SOURCE: RAND Arroyo analysis of TAPDB-AE and ReARMM data.

the general lack of pattern within any of the ReARMM phases (with some exceptions, outlined in Appendix A), as well as the relatively broad distribution of gain and loss rates for units. Each dot in the figures illustrates a unit's gain or loss rate for that month, with the color indicating the phase of ReARMM.

The results illuminate a few noteworthy phenomena. First, the general loss and gain rates do not, in most cases, decrease or increase within a given phase (again, specific details are provided in Appendix A). This result could be interpreted as a mixed outcome. This generally uniform distribution of gains and losses within a phase is presumably better than losing or gaining more personnel toward the end of some phases but less ideal than more of the turnover in an earlier portion of a phase. Secondly, while the average monthly unit gain and loss rate is generally close to the 2–3 percent HRC goal,[9] there are significant outliers, with many units gaining or losing over 5 percent. While we did not explore the specific unit turnover rates, it is certainly possible that those units with high turnover in one month are not necessarily immune from high turnover in the next. Exploring which units see greater turnover is an area where future research could lend insights into which units may need greater mitigation or policy levers.

Dimension 7 Results: Associations Between Retention and People First Metrics

Dimension 7 explored the degree to which MOS match (Dimension 1) and assignment considerations (Dimension 2)[10] were associated with retention.[11] For each variable of interest, we ran a regression model controlling

---

[9] As noted, ReARMM is too recent for us to perform a longitudinal analysis that assesses whether it has had impacts on reenlistment rates.

[10] Unlike Dimensions 1 and 2, which characterize soldier-level factors, Dimensions 3, 5, and 6 are unit-level dimensions. Dimension 4 was not analyzed at the individual level because most soldiers had missing values for that dimension.

[11] For retention analysis, we analyzed the population of soldiers in FORSCOM brigades that had a reenlistment date between January 2020 and September 2022. Individuals who completed their term and reenlisted as of September 2022 (the end of our period of analysis) are categorized as "retained." Individuals who did not reenlist or who separated prior to their ETS date are categorized as "non-retained."

for gender, race, whether the soldier was assigned to a combat MOS, and soldier pay grade. To best isolate effects of the control variables, the model considered interaction terms between gender and the remaining variables (i.e., the model was fully interacted with gender). Further details of the setup and results of the regression models can be found in Appendix A.

Results from the general model (that is, the model that includes only control variables, without including a specific dimension variable of interest) give context to the remainder of this section. Consistent with other retention research, the general model finds that retention was more likely among male soldiers than female soldiers (p = 0.000), among nonwhite soldiers than white soldiers (p = 0.000), and among soldiers in pay grades E5 and higher (p = 0.000). We also found that retention in our study units was more likely among soldiers that did not have a combat MOS than those that did (p = 0.000). The race effect was larger for women than for men (p = 0.003); the grade effect was larger for men (p = 0.000). The effect of combat MOS was not significantly different between men and women (p = 0.856).

In Dimension 1, we analyzed three degrees of MOS match. A first-degree match considers whether a soldier's MOS matches their duty MOS. A second-degree match considers MOS and SQI, and a third-degree match considers MOS, SQI, and ASI. Results from the regression analysis indicated no significant association between first-degree and second-degree matches and retention (p = 0.124 and p = 0.249, respectively). That is, there is no statistically significant evidence to indicate that a soldier whose MOS and SQI match their duty assignment was more or less likely to be retained than a soldier with mismatching MOS and SQI. However, when ASI was considered in addition to MOS and SQI, soldiers with a third-degree MOS match were 7 percent more likely to be retained than others.[12] These results are marginally significant, with a p-value of 0.056, and did not differ significantly by gender. To put this result in the context of ReARMM, Dimension 2 analysis found slightly higher levels of third-degree MOS match during the mission–PTDO ReARMM phase than during modernization.

---

[12] We note that the risk ratios for the three degrees of MOS match do not differ significantly from each other. This suggests that all three match levels may matter, though only the third-degree match results that have a lower overall match rate and lower standard error of estimate approach statistical significance. See Tables A.26–A.28.

Dimension 2 analyzed four types of soldier assignment considerations: medical, conduct, standards, and family considerations. Results indicate that retention was 60 percent less likely if a female soldier had a medical consideration (p = 0.000). The medical-gender interaction variable indicates that this effect was greater for women than for men by about 10 percentage points (p = 0.000). For context within ReARMM, Dimension 2 found that medical considerations were more prevalent in mission–PTDO than in the modernization phase.

Results also indicate that retention was 29 percent less likely if a soldier had a conduct consideration (p = 0.000). There were no significant differences in this effect by gender (p = 0.193). Dimension 2 found conduct considerations to be more prevalent during training than during modernization.

The analysis found no significant association between standards considerations and retention; that is, a soldier who had a standards consideration was no more or less likely to be retained than a soldier who did not have a standards consideration. It is possible that, compared to medical, conduct, and family considerations, a standards consideration is more easily remedied (for example, a soldier can lose weight to meet weight standards or pass an education course that was previously incomplete). Indeed, Dimension 2 found that standards considerations become less prevalent over the ReARMM phases.

Finally, results indicate that associations between family considerations and retention depend fundamentally on the soldier's gender. Women with a family consideration were 26 percent *less* likely to be retained (p = 0.000); however, men with a family consideration were 8 percent *more* likely to be retained (p = 0.000). The Army may want to consider further research to explore reasons for this difference. Within the ReARMM context, Dimension 2 found family considerations to be more prevalent during the training phase than during modernization.

Conclusion

This chapter analyzed a set of quantitative analysis dimensions to assess friction points between People First goals and elements of ReARMM. While some results do not point to obvious frictions within specific ReARMM cycles, many provide insight into potential actions to mitigate specific frictions. The quantitative analysis leads to several findings that complement and provide analytical backing to the findings from the interviews, scenario results, and review of the literature. We outline those findings and specific recommendations in the next chapter.

CHAPTER 5

Conclusions and Recommendations

The Army operates in an inherently unpredictable environment. International crises, domestic demands, and the need to practice dealing with the unexpected all contribute to the fact that predictability, as a goal, can only be provided up to a certain extent. Whether conducting ARFORGEN, SRM, or ReARMM, soldiers understand (or should) that the training schedule can only provide imperfect certainty.

To meet this uncertainty, the Army needs to maintain a constant flow of ready units, whether the unit will actually deploy for an ongoing mission, the unit is prepared for a specific on-call mission, or the unit is just at a high state of readiness and available for the unexpected. Units cannot stay in this ready state forever. People rotate in and out, and therefore the unit needs to go back through training cycles to build the teamwork and cohesion demanded by Army operations. Old equipment must be traded for new, and soldiers and leaders trained on how to employ and sustain that new equipment and how to integrate into their operations. At a macro level, ReARMM can be an effective process for managing this allocation of units to various mission lines or profiles and is central to the Army's ability to meet its operational mission commitments.

Units are made up of people. The Army has recognized that it has some people challenges—the Fort Hood report being a significant indicator of those challenges. The Army People Strategy and the Army's talent management program are two key aspects to an Army imperative to put "People First." Work-life balance, the Army senior leadership's vision of "cohesive teams" of soldiers who truly understand and accept each other, and granting more power to soldiers in management of their careers are all happening at a micro level (i.e., within units). Each of these aspects of the emerging Army culture and practice take time away from the already overburdened and crowded demands for soldier, team, and unit training. In other words, there will be friction between the unit readiness goals and People First.

Our research was designed to understand that friction better and see whether (and how) it could be mitigated. What we found is that macro-level adjustments to synchronizing ReARMM are probably not the answer. Within the constraints, incentives, and goals of their own macro systems, both the Army's Personnel and Modernization enterprises have little incentive and few degrees of freedom to interact differently with ReARMM. Moreover, we also did not find that such macro adjustments are necessary. Instead, our findings suggest that while there may be some specific areas of personnel management that could be adjusted, the best set of responses to addressing the friction we found may come from communications and culture changes.

The most important piece of this is that the Army is not communicating consistently about its priorities from the highest level to the lowest. In addition, it has not aligned its incentive structure to reward meeting People First goals, even though the senior leaders state clearly and publicly that that is their top priority. This kind of culture change and incentive change is difficult and will take time to work through the system. However, absent movement in this direction, it appears obvious that the friction will remain. We do not attempt to answer the question of whether the current friction is at an acceptable level, but we do recognize both that it exists and that there are options to mitigate its effects, if not always its presence.

Detailed Findings

Our findings fall into several categories, each of which highlights a different kind of friction and therefore generates the need for a different kind of response. The first two findings apply to all Army units subject to ReARMM and revolve around (1) communications and messaging and (2) priorities and incentives. The next finding represents an area we already know that the Army is beginning to address: ReARMM (at the time we investigated it) is a BCT-centric process that does not fit with the mission demand profile of many Army units. We provide a set of findings that relate to how ReARMM intersects with the Army's personnel and modernization enterprises. Last, we provide findings derived from our analysis of Army personnel data and our attempts to better measure and predict friction.

Findings from Qualitative Analysis of Interviews and Policy

At the level of the Secretary of the Army, the CSA, the Sergeant Major of the Army, and the FORSCOM commander, the need to put people first and to accept some short-term readiness risk in order to build toward long-term health of the force is clear. Three of the Secretary's top six priorities are clearly drawn from the People First approach. The CSA has said that people are his number one priority.[1] Sgt. Maj. of the Army Michael Grinston emphasized the need for soldiers to have time off and for units to have white space on their calendars.[2]

However, this is not the message being heard at the troop unit level. Two interventions are impacting how soldiers actually behave in ways that may not be aligned with the Army senior leader priorities. The first of these is that the Army's message of People First is diluted or even missing from prioritization and guidance documents at unit level. The further down the chain of command, the more dilution seems to be taking place. Instead, a more traditional focus on near-term tactical unit readiness seems to be the primary goal of units. Moreover, the incentive structures in place for unit leaders reinforce this shift from Army to unit-level priorities.

Adding to the challenges faced by unit leaders is that the modernization and personnel systems, which both place demands—often unpredictable ones—on units, are not incentivized and are not flexible in ways that can materially mitigate the challenges of friction. HRC has a set of broader personnel goals and constraints that mitigate against the kinds of tailored personnel assignment and management processes that would be necessary to fix friction. As a result, management of that friction falls on unit personnel managers and on unit leaders. Similarly, program executive offices

---

[1]   Michelle Tan, "Putting People First: McConville Looks to Revolutionize How Soldiers Serve," Association of the United States Army, October 3, 2019.

[2]   Haley Britzky, "The Sgt. Maj. of the Army Wants Leaders to Stop Scheduling Training Just for the Sake of It," *Task & Purpose*, October 18, 2021.

and project managers have schedules to meet and costs that are incurred as those schedules slip. Army processes for programming material change in units are also prone to rapid changes that respond to more macro Army funding and operational challenges. In short, synchronization and optimization are not possible at a macro level, though some marginal changes may be possible.

We identified that there are potentially some cultural and generational challenges: Does the Army really need People First? Isn't that what leadership is already supposed to be about? Is the goal more time off? Or is it more meaningful training? These combine with perceived challenges in the understanding and execution of Army training management processes. Poorly synchronized training resources and senior leader capture of unit-level white space all contribute to a sense of loss of control or predictability that create leader dilemmas ("What do I tell my soldiers? How can I keep my promises?") and create false expectations among the troops.

Of lesser note, we found an overwhelming belief that ReARMM is a BCT-focused process that mostly applies to ABCTs (and maybe combat aviation brigades) but has little relevance or utility for brigades with fewer modernization needs or with support requirements that defy ReARMM phasing. We do note that the Army is currently developing and publishing ReARMM updates that are designed to tailor the process more to non-BCT units.

Findings from Quantitative Analysis of Personnel and ReARMM Data

The quantitative analysis leads to several findings that complement and provide analytical backing to the findings from the interviews, scenario results, and review of the literature. In addition to resource constraints related to data collection and processing, ReARMM is too new (circa FY 2020) for us to do effective longitudinal analysis of its effects on certain People First initiatives. While we did find patterns that were statistically significant, how those patterns change over time and among different units' ReARMM cycles was not something the data currently allow but is something the Army should consider investigating as more data become available.

The level of mismatch between soldiers' primary MOS (or secondary or additional MOS, if applicable) and their duty MOS was generally small (less than 4 percent of personnel) but was statistically larger for the deployment phase of ReARMM for FORSCOM BCT units. This result appears to support some concern expressed in the interviews about mismatch, especially since the mismatch increases statistically significantly during deployment, which can be recognized as a factor for cohesive teams and talent management.

Results from assignment considerations analyzed in Dimension 2 indicate a few areas on which the Army may wish to focus. For instance, while assignment considerations for family or conduct appeared slightly more frequently during the training phase of ReARMM, they were not significantly higher during deployment. This result, combined with the result from Dimension 3 that fill rates were not statistically lower during deployment, may imply that an increased number of soldiers with assignment considerations for family or conduct does not result in lower deployment fill rates. For FORSCOM BCTs, this result may not be surprising, as manning guidance prioritizes fill rates of these units.[3] However, conducting this analysis for other types of units may yield results that are less intuitive and more insightful. Commander awareness of these trends may help ease these stressors leading up to the deployment phase.

Examining duty location preference compared to soldier assignments showed a modest decrease in preference match during mission–deployment. This result likely reflects the need to prioritize fill rates during this phase. The decrease, while statistically significant, was modest in absolute terms. Perhaps more significant is that the overall preference to actual location match was low—about 25 percent for those that indicated a preference—and may reflect some degree of work-life balance concerns. Should this have significant impacts on retention and recruitment, efforts to reconcile this mismatch may be of interest.[4]

---

[3] HRC/G1, interviews with the authors, June 1, 2022.

[4] A caveat to these results is the fact that, for the study population (enlisted soldiers in FORSCOM BCTs), only around 30 percent of individuals have a location preference listed in the TAPDB-AE data. Whether this is because the soldiers were not asked their preference, whether they did not have a preference, or whether there was some other reason is unclear.

We found that the overall level of attendance of resident training courses during the ReARMM cycle was very low. Although there were a few differences in the attendance rate over the phases, they were considerably smaller than the low overall level in absolute terms.

We found that turnover in units through the ReARMM cycle paints a nuanced picture. The overall monthly gains and losses for units were generally in the HRC aggregate 2–3 percent target range. In addition, there did not seem to be increased gains and losses during less desirable times, such as toward the end of a training phase. On the other hand, there were significant statistical outliers, with many above 5 percent and several in the 7–8 percent range. These units may reflect the anecdotal criticisms of units experiencing high turnover. In addition, while there were not strong trends of higher turnover during undesirable time periods, there was similarly little evidence of lower turnover. It would seem that planning for more turnover during less disruptive periods in a phase (e.g., early in training) does not widely occur.

While friction is clearly happening and was suggested by the interviews, it is possible that the level of incidents (as suggested by the quantitative data) may be less than commonly believed. For most things we explored, the data suggest that most of the time, for most units, the problems do not appear to have been of a large magnitude. These quantitative results may be useful as FORSCOM explores how to address and manage change oriented toward minimizing friction.

Some of the Dimension 1 and Dimension 2 People First metrics are associated with retention. A soldier with a third-degree MOS match (i.e., MOS, SQI, and ASI all match their duty assignment) was slightly more likely to be retained, although this effect was only marginally significant. Several assignment considerations were also associated with retention: Soldiers with conduct considerations were much less likely to be retained than soldiers without those considerations. Soldiers with medical considerations also were much less likely to be retained, but the effect of medical considerations was greater for women than for men. Women with family considerations were considerably less likely to be retained, while men with family considerations were modestly more likely to be retained.[5]

Detailed Recommendations

Our recommendations follow from our findings. The Army needs to find ways to clarify its message and to ensure that the message conveys from top to bottom. Improving the processes of training management (and training people on it) may also be helpful, but of more consequence is the need to find ways to shift Army incentives by quantifying or better defining People First goals in ways that both allow Army leaders to identify who is being successful at implementing People First and to then document that success in ways that materially affect careers. We recognize that the Army will always have a certain amount of unpredictability and uncertainty—it is the nature of the Army's mission. However, the Army needs to consider providing training and resources to junior (and senior) leaders on how to communicate uncertainty in ways that can help soldiers manage expectations and reduce feelings of cognitive dissonance between what they hear from the Army's senior leadership and what they experience at unit level.

The quantitative analyses serve as examples of the types of considerations the Army can investigate to illuminate friction points between ReARMM and People First initiatives. Our recommendations here are somewhat more nebulous, in that they require Army decisions about which analytics to pursue and how to use them, rather than immediately actionable recommendations, like those above. These initial personnel data analyses investigated FORSCOM BCTs. To generate a clearer picture of forcewide implications of ReARMM, the Army should conduct similar analyses for other types of units, as ReARMM continues to be implemented Army-wide. Because the implementation of ReARMM in BCT units is so recent, the Army should continue to conduct these types of analyses as ReARMM implementation becomes more mature. Finally, these analyses often tell opposing stories about friction points between ReARMM and People First initiatives. Some

---

[5] There were no statistically significant associations between retention and first-degree MOS match, second-degree MOS match, or standards considerations.

analyses—such as unit turnover—indicate little or no friction points at the aggregate level. However, the data also indicate that despite predictability at the aggregate level, individual units, grades, MOSs, or soldiers may experience outlier events that can be very disruptive. As a result, the Army should explore and balance both the aggregate trends as well as the stress points for outliers.

Here we recap, at a more specific level, the recommendations summarized above. Of note, many of these intersect with each other, and we try to highlight some of the key connections. These recommendations also incorporate some of the best practices we identified from the literature on expectation management.

Inculcate Initiatives into Army Culture and Communications

Qualitative analysis provided us an opportunity to understand how soldiers and leaders experience ReARMM, through a lens of People First, and to better understand what friction points may emerge during implementation. Insights we captured also bolstered our thinking on potential mitigation strategies that might ease the transition. Some of the discussion below may reinforce this particular recommendation, but they do not substitute for an Army commitment to culture change and to clarity and consistency in communications. The People First programs, the Army People Strategy, and the Army's talent management initiatives have, we believe, already started to impact culture. The Army should find ways to sustain or increase the rate of change and ensure that it becomes institutionalized in its schools, performance measures, and culture.

Clearly and Consistently Message People First Principles Across the Force

It will be difficult to achieve People First objectives if the Army lacks a clear and consistent approach to implementing guidance in practice. And because People First will inevitably mean different things to different people depending on their individual priorities, Army leaders will want to communicate concrete examples of how to demonstrate that commitment. Our analysis suggests that the Army should identify more deliberate and engaging ways to socialize these concepts.

Another way to help communicate and build consistency in messaging may be to conduct role-playing training similar to the workshop we conducted. Our workshop provided interesting insights into the potential power of persuasion. Asking volunteers to concentrate on ReARMM and People First challenges allowed this cohort of officers to empathize with their fellow soldiers (and, in some cases, relive their own familiar dilemmas) and collectively explore potential mitigation strategies that favored People First goals while still meeting ReARMM intent. Providing a similar opportunity (or multiple opportunities) in both classrooms and units can also help build this recommended consistency in messaging and can reinforce the desired culture.

Ensure That Communication Is a Two-Way Street

The Army should develop and implement a way to rapidly gather feedback on friction issues from unit-level leaders. While conversations taking place at higher echelons will be important to actioning initiatives addressing friction challenges in the months ahead, our research suggests it will be critical to include operational-level units at the table to ensure that (1) Army leaders have a clear understanding of how concepts function (or struggle to) in practice; and (2) leaders at lower echelons of command are provided clear and consistent guidance on how to execute.

Develop Clear and Measurable Metrics to Assess Army Leaders' Commitments to People First

We recommend that the Army find ways to include People First–related metrics or comments as mandatory inclusions in OERs/NCOERs and in counseling sessions. Command climate surveys are another mechanism to consider, though developing the right metrics to help decipher trends toward or away from ReARMM and People First prerogatives will be critical. For the Army to do this, it must develop clear and measurable metrics or standards that can be translated into this kind of performance feedback.

Identify Opportunities to Better Synch Initiatives with Existing Processes; Tinkering at the Margins Can Make a Difference

The Army should identify specific opportunities, rather than broad systemic ones, where it can synchronize its personnel processes to a unit's place in the ReARMM cycle to enhance predictability. One example relates to PCS and ETS policies. Commanders lack flexibility in working with soldiers who are leaving the unit due to a PCS or ETS shortly after return from deployments, and changes allowing commanders greater flexibility to address soldiers' needs on a case-by-case basis would be a positive change. For example, interviews suggested aligning PME school dates to ReARMM cycles and better aligning manning processes to unit needs, particularly in the modernization phase of ReARMM, when demand for soldiers with specialty MOSs is critical. HRC identifying ways to ensure that transitions in and out of a unit primarily occur during the modernization period and that unit leaders have the ability to stabilize soldiers in this phase as needed would also promote unit cohesion and readiness but may not fit well with overall HRC and Army goals. Our analysis did not go deep enough to recommend the right balance between these goals, but we believe that the Army should look to see if a rebalance is possible for specific issues or jobs, even if not for the whole of the force.

Communicate with an Understanding of Expectation Management

One of ReARMM's goals is to build predictability. And, at a macro level, it can often do so. As we have noted several times, the Army has a mission for which unpredictability is a defining characteristic. Nevertheless, the Army attempts to communicate expectations to its soldiers in ways that allow them to plan for both themselves and their families. Our literature review of expectation management—the concept of how to communicate both certainty and uncertainty to minimize the negative impacts of unexpected changes—suggests that the Army may want to consider looking at that literature and incorporating elements into its leader training programs. The Army should teach junior leaders to communicate the uncertainty and to communicate in ways that temper their soldiers' expectations. One way to get at this is to consider including a program of instruction in junior-level training courses (Company Command, Noncommissioned Officer Education System) on expectation management and communication of uncertainty.

Consider Further Quantitative Analyses for Mitigating Friction Points

Based on our quantitative analyses, we have the following recommendations:

- These analyses serve as examples of the types of considerations the Army can investigate to illuminate and better understand friction points between ReARMM and People First initiatives. These initial analyses investigated FORSCOM BCTs. To generate a clearer picture of forcewide implications of ReARMM, the Army should conduct similar analyses for other types of units as ReARMM continues to be implemented Army-wide.
- Similarly, because the implementation of ReARMM in BCT units is so recent, these results are based on units in flux (i.e., units converting to ReARMM) rather than units in stasis (i.e., units that have been through several stable ReARMM cycles). To further understand the implications of ReARMM and its friction points with People First initiatives, the Army should continue to conduct these types of analyses as ReARMM implementation becomes more mature.
- Finally, two opposing stories can be told about friction points between ReARMM and People First initiatives. Some analyses—such as unit turnover—indicate little or no friction points at the aggregate level. However, the data also indicate that despite predictability at the aggregate level, individual units, grades, MOSs, or soldiers may experience outlier events that can be very disruptive. When considering friction points and policy alternatives, the Army should explore and balance both the aggregate trends and the stress points for outliers. Prescriptive analytics may be able to identify ways to mitigate those disruptions for units with high or unpredictable turnover rates.

Conclusion

We began this research thinking that the challenge would be how to better synchronize macro Army processes. We quickly found that the friction is more about people and culture than process and calendar. The Army leadership clearly understands the inherent tensions between near-term unit readiness and long-term health of the force. The Army leadership is, we think appropriately, focused on culture changes that address unwanted behavior in the ranks. And they understand that unit leaders, down to the squad level, need time and resources and additional supporting programs in order to focus on and implement that change. However, Army leadership—and unit leaders at all levels—are also focused on the next mission. The nation's demand signal for Army capabilities suggests that many units will have to be ready—and quickly—for both the known missions and the contingent ones. So, friction is inevitable.

Management of that friction will not be accomplished by process change per se. It requires clear communication, consistent messaging, priorities aligned with incentives, and, perhaps, some additional training in certain discrete areas (e.g., training management and expectation management). Our research does not offer much in the way of specifics about how to do this. It does point to the areas of focus. We demonstrate a role that detailed quantitative research can play in addressing a few key points of friction. But, as some of our interviewees mentioned, much of this is "Leadership 101" and is well within the ability of the Army to address, if it can clearly tell the leaders what is expected and then hold them to those expectations.

Our final word is about the nature of friction itself. While friction is clearly occurring and Army leaders are clearly being challenged with dilemmas relating to balancing ReARMM and People First, the magnitude and impact of friction is not clear. Some friction in any system is inevitable. All friction is worth being addressed. However, in another balancing challenge, the Army may want to decide where to invest its energy. If friction cannot be eliminated by systemic change, maybe it can be mitigated by cultural adaptation. Our recommendation would be to focus on the communication and incentive aspects of the problem, while acknowledging that, time and resources being available, there is room for some micro systemic adjustments (for which we have provided specific ideas in this chapter).

Regression Statistics

This appendix contains detailed statistical results from the quantitative analysis described in Chapter 4, "Quantitative Analysis of People First and ReARMM Friction Points." The quantitative analysis covers seven dimensions of ReARMM and People First friction points, which are summarized here. For Dimensions 1–6, the results tables detail the statistical means and regression coefficients of the dimension variables, the differences (deltas) for the variables from the modernization phase level, standard errors of the coefficients, t-statistics, and p-values for each of the predictors (here, ReARMM phases). The modernization phase was used as the base for each of the regressions; therefore, modernization shows a statistical mean only, while the remaining regression variables indicate whether and how the other phases differ from that statistical mean. Results are considered statistically significant (Y) when p-value < 0.050, marginally significant (M) when 0.050 ≤ p-value < 0.100, and not significant (N) when the p-value ≥ 0.100.

Dimension 1: Match Between Soldiers' MOS and Duty MOS over ReARMM

This dimension assessed within each FORSCOM unit the magnitude of the mismatch between a soldier's MOS and the soldier's assigned duty MOS over the ReARMM cycle. This comparison also includes SQI and ASI. Table A.1 lists the TAPDB AE variables considered in this analysis, and Table A.2 indicates the conditions required for an individual's record to be counted as a "match."

TABLE A.1

TAPDB-AE Variables for MOS Match Analyses

Variable Name	TAPDB-AE Variable
Duty MOS	DYMOSE
Primary MOS	PMOSEN
Secondary MOS	SMOSEN
Additional MOS	AMOSEN
Duty SQI	DYSQIE
SQI	SQIEN
Control SQI	CTSQIE
First Duty SQI	FDASIE
ASI	ASIEN
Control ASI	CTASIE

SOURCE: RAND Arroyo Center analysis of TAPDB-AE and ReARMM data.

TABLE A.2

Matching Conditions for MOS, SQI, and ASI Matches

Match Type	Matching Conditions
MOS match	DYMOSE = PMOSEN or SMOSEN or AMOSEN
MOS and SQI match	DYMOSE = PMOSEN or SMOSEN or AMOSEN and DYSQIE = SQIEN or CTSQIE; or DYSQIE = "O"; or DYSQIE is unpopulated
MOS, SQI, and ASI match	DYMOSE = PMOSEN or SMOSEN or AMOSEN and DYSQIE = SQIEN or CTSQIE; or DYSQIE = "O"; or DYSQIE is unpopulated and FDASIE = ASIEN or CTASIE; or FDASIE = "00"; or FDASIE is unpopulated

SOURCE: RAND Arroyo Center analysis of TAPDB-AE and ReARMM data.

Tables A.3, A.4, and A.5 detail the statistical results of the MOS match regressions and note whether the differences between modernization and other ReARMM phases are statistically significant.

TABLE A.3

Regression Statistics for ReARMM Phase and MOS Match

ReARMM Phase	Mean	Delta	Standard Error	T	p>|t|	Significant?
Modernization	0.967912	--	--	--	--	--
Training	0.967102	−0.0008107	0.0012773	−0.63	0.526	N
Mission–PTDO	0.967332	−0.0005804	0.0015414	−0.38	0.707	N
Mission–Deploy	0.960812	−0.0071005	0.0014407	−4.93	0.000	Y

SOURCE: RAND Arroyo Center analysis of TAPDB-AE and ReARMM data. N = 874.

TABLE A.4

Regression Statistics for ReARMM Phase and MOS Match and SQI Match

ReARMM Phase	Mean	Delta	Standard Error	T	p>|t|	Significant?
Modernization	0.928776	--	--	--	--	--
Training	0.923982	−0.0047944	0.0035904	−1.34	0.182	N
Mission–PTDO	0.927224	−0.0015523	0.0043326	−0.36	0.72	N
Mission–Deploy	0.9125	−0.0162758	0.0040497	−4.02	0	Y

SOURCE: RAND Arroyo Center analysis of TAPDB-AE and ReARMM data. N = 874.

TABLE A.5

Regression Statistics for ReARMM Phase and MOS Match, SQI Match, and ASI Match

ReARMM Phase	Mean	Delta	Standard Error	T	p>|t|	Significant?
Modernization	0.809956	--	--	--	--	--
Training	0.809585	−0.0003713	0.0048732	−0.08	0.939	N
Mission–PTDO	0.823233	0.0132775	0.0058807	2.26	0.024	Y
Mission–Deploy	0.815729	0.0057729	0.0054966	1.05	0.294	N

SOURCE: RAND Arroyo Center analysis of TAPDB-AE and ReARMM data. N = 874.

Dimension 2: Emergence of Specific Soldier Characteristics over ReARMM

We assessed whether the number of soldiers within a FORSCOM unit affected by special considerations (including assignment considerations and delay or deferral of assignment) related to medical, failure to meet standards, conduct, and family varied over the unit's ReARMM cycle. Table A.6 lists the TAPDB-AE variables considered in this analysis.

Tables A.7, A.8, A.9, and A.10 detail the statistical results of these regressions and whether the differences between modernization and other ReARMM phases are statistically significant.

TABLE A.6

TAPDB-AE Variables for Soldier Characteristics

Characteristic	Variable Name	TAPDB-AE Variable	Example Entries
Medical	Assignment Consideration	ASCO	Physical limitations
Medical	Promotion Ineligibility Reason	PRMINL	Medical disqualification
Medical	PULHES Ratings	PHCP, UPEX, LOEX, VIS, HRNG, PSYC	Significant limitations, below retention standards
Medical	Major Personnel Action Reason	MPARSN	Pregnancy, disability, personality disorder
Medical	Deletion/Deferment Justification Code	DDRQCD	Medical/dental treatment
Medical	Assignment Eligibility and Availability	AEA	Pending medical evaluation
Medical	Military Rank Change Reason	RNKCRS	Disability evaluation system
Medical	Reassignment Reason	RSGRSN	Medically unfit for retention, pregnancy, release from medical hold detachment
Medical	Immediate Enlistment Prohibition	IMREPR	Pending medical evaluation

Table A.6—Continued

Characteristic	Variable Name	TAPDB-AE Variable	Example Entries
Medical	Deletion/Deferment Reason	DLDFRN	Medical/dental treatment, pregnant, pending medical evaluation
Medical	Enlistment Prohibition Waiver	ENPRWV	Medical discharge
Standards	Suspension of Favorable Personnel Action	SFPARS	Army physical fitness test failure
Standards	Enlistment Prohibition Ineligibility Reason	PRMINL	Does not meet Noncommissioned Officer Education System requirement, does not meet security requirement, failed to meet physical fitness or weight standards
Standards	Major Personnel Action Reason	MPRSN	Failure to maintain Army weight control standards, unsatisfactory performance, physical standards
Standards	Military Rank Change Reason	RNKCRS	Inefficiency, failure to complete schooling
Standards	Reassignment Reason	RSGRSN	Loss of qualification
Standards	Immediate Enlistment Prohibition	IMREPR	Denied retention by Secretary of the Army— commander quality, physical readiness, loss of qualification, weight
Standards	Deletion/Deferment Reason	DLDFRN	Airborne training failure, academic failure
Conduct	Suspension of Favorable Personnel Action	SFPARS	Adverse action, commander's investigation, law enforcement investigation, drug or alcohol abuse adverse action
Conduct	Reenlistment Eligibility	REELRA	Ineligible without a waiver due to time lost

Table A.6—Continued

Characteristic	Variable Name	TAPDB-AE Variable	Example Entries
Conduct	Immediate Enlistment Prohibition	IMREPR	Lost time, adverse action, court-martial conviction
Family	Major Personnel Action Reason	MPARSN	Pregnancy, parenthood
Family	Deletion/Deferment Justification Code	DDRQCD	Compassionate, extreme family problem
Family	Reassignment Reason	RSGRSN	Compassionate, joint domicile, exceptional family member, pregnancy
Family	Deletion/Deferment Reason	DLDFRN	Compassionate, pregnancy, extreme family problem, exceptional family member
Family	Enlistment Prohibition Waiver	ENPRWV	Dependents, sole parent, hardship, sole survivor

SOURCE: RAND Arroyo Center analysis of TAPDB-AE and ReARMM data.

TABLE A.7

Regression Statistics for ReARMM Phase and Medical Variables

ReARMM Phase	Mean	Delta	Standard Error	T	p>|t|	Significant?
Modernization	0.043365	--	--	--	--	--
Training	0.045120	0.0017542	0.0010208	1.72	0.086	M
Mission–PTDO	0.046982	0.0036167	0.0012319	2.94	0.003	Y
Mission–Deploy	0.042871	−0.0004940	0.0011514	−0.43	0.668	N

SOURCE: RAND Arroyo Center analysis of TAPDB-AE and ReARMM data. N = 874.

NOTE: M = marginally significant.

TABLE A.8

Regression Statistics for ReARMM Phase and Standards Variables

ReARMM Phase	Mean	Delta	Standard Error	T	p>|t|	Significant?
Modernization	0.030791	--	--	--	--	--
Training	0.028602	-0.0021886	0.0007821	-2.80	0.005	Y
Mission–PTDO	0.024072	-0.0067191	0.0009437	-7.12	0.000	Y
Mission–Deploy	0.024650	-0.0061404	0.0008821	-6.96	0.000	Y

SOURCE: RAND Arroyo Center analysis of TAPDB-AE and ReARMM data. N = 874.

TABLE A.9

Regression Statistics for ReARMM Phase and Conduct Variables

ReARMM Phase	Mean	Delta	Standard Error	T	p>|t|	Significant?
Modernization	0.060529	--	--	--	--	--
Training	0.064930	0.0044001	0.0012235	3.60	0.000	Y
Mission–PTDO	0.063045	0.0025158	0.0014764	1.70	0.089	M
Mission–Deploy	0.061049	0.0005193	0.0013800	0.38	0.707	N

SOURCE: RAND Arroyo Center analysis of TAPDB-AE and ReARMM data. N = 874.

TABLE A.10

Regression Statistics for ReARMM Phase and Family Variables

ReARMM Phase	Mean	Delta	Standard Error	T	p>|t|	Significant?
Modernization	0.018850	--	--	--	--	--
Training	0.019595	0.0007446	0.0002641	2.82	0.005	Y
Mission–PTDO	0.018793	-0.0000575	0.0003187	-0.18	0.857	N
Mission–Deploy	0.018364	-0.0004857	0.0002978	-1.63	0.103	N

SOURCE: RAND Arroyo Center analysis of TAPDB-AE and ReARMM data. N = 874.

Dimension 3: Differences Between a Unit's MTOE and the Number of Assigned and Available Soldiers over ReARMM

Building off the analysis in Dimension 2, if certain assignment considerations appear more frequently in certain phases of ReARMM, we aim to measure their impacts—if any—on personnel shortfalls during those phases. We analyze the differences between the number of soldiers specified in the MTOE for a unit and the assigned soldiers over the ReARMM cycle to see whether differences are greater in certain phases, notably ones associated with higher degrees of soldiers with special assignment considerations. We performed analyses on the overall brigade fill ratio and broke down the fill ratios by grades: E4 and below, E5, E6, E7, E8, and E9. Tables A.11 through A.17 detail the statistical results of these regressions and whether the differences between modernization and other ReARMM phases are statistically significant.

TABLE A.11

Regression Statistics for ReARMM Phase and Overall MTOE Fill Rate

ReARMM Phase	Mean	Delta	Standard Error	T	p>|t|	Significant?
Modernization	1.049217	--	--	--	--	--
Training	1.047380	−0.0018368	0.0052327	−0.35	0.726	N
Mission–PTDO	1.069788	0.0205709	0.0063144	3.26	0.001	Y
Mission–Deploy	1.052253	0.0030357	0.0059021	0.51	0.607	N

SOURCE: RAND Arroyo Center analysis of TAPDB-AE and ReARMM data. N = 5244.

TABLE A.12

Regression Statistics for ReARMM Phase and E1 Through E4 MTOE Fill Rate

ReARMM Phase	Mean	Delta	Standard Error	T	p>|t|	Significant?
Modernization	1.191969	--	--	--	--	--
Training	1.200463	0.0084936	0.0058488	1.45	0.147	N
Mission–PTDO	1.203740	0.0117706	0.0070580	1.67	0.096	M
Mission–Deploy	1.215912	0.0239431	0.0065971	3.63	0.000	Y

SOURCE: RAND Arroyo Center analysis of TAPDB-AE and ReARMM data. N = 874.

TABLE A.13

Regression Statistics for ReARMM Phase and E5 MTOE Fill Rate

ReARMM Phase	Mean	Delta	Standard Error	T	p>|t|	Significant?
Modernization	1.065306	--	--	--	--	--
Training	1.083907	0.0186007	0.0151898	1.22	0.221	N
Mission–PTDO	1.084625	0.0193192	0.0183300	1.05	0.292	N
Mission–Deploy	1.007757	−0.0575494	0.0171330	−3.36	0.001	Y

SOURCE: RAND Arroyo Center analysis of TAPDB-AE and ReARMM data. N = 874.

TABLE A.14

Regression Statistics for ReARMM Phase and E6 MTOE Fill Rate

ReARMM Phase	Mean	Delta	Standard Error	T	p>|t|	Significant?
Modernization	1.099716	--	--	--	--	--
Training	1.084275	−0.0154413	0.0091176	−1.69	0.091	M
Mission–PTDO	1.106843	0.0071270	0.0110025	0.65	0.517	N
Mission–Deploy	1.065166	−0.0345505	0.0102840	−3.36	0.001	Y

SOURCE: RAND Arroyo Center analysis of TAPDB-AE and ReARMM data. N = 874.

TABLE A.15

Regression Statistics for ReARMM Phase and E7 MTOE Fill Rate

ReARMM Phase	Mean	Delta	Standard Error	T	p>|t|	Significant?
Modernization	1.000481	--	--	--	--	--
Training	1.009795	0.0093135	0.0061786	1.51	0.132	N
Mission–PTDO	1.045965	0.0454844	0.0074559	6.10	0.000	Y
Mission–Deploy	1.035676	0.0351952	0.006969	5.05	0.000	Y

SOURCE: RAND Arroyo Center analysis of TAPDB-AE and ReARMM data. N = 874.

TABLE A.16

Regression Statistics for ReARMM Phase and E8 MTOE Fill Rate

ReARMM Phase	Mean	Delta	Standard Error	T	p>|t|	Significant?
Modernization	0.956314	--	--	--	--	--
Training	0.956141	−0.0001734	0.0117222	−0.01	0.988	N
Mission–PTDO	1.002411	0.0460974	0.0141455	3.26	0.001	Y
Mission–Deploy	0.978376	0.0220616	0.0132218	1.67	0.096	M

SOURCE: RAND Arroyo Center analysis of TAPDB-AE and ReARMM data. N = 874.

TABLE A.17

Regression Statistics for ReARMM Phase and E9 MTOE Fill Rate

ReARMM Phase	Mean	Delta	Standard Error	T	p>|t|	Significant?
Modernization	0.981518	--	--	--	--	--
Training	0.949704	−0.0318139	0.013275	−2.40	0.017	Y
Mission–PTDO	0.975145	−0.0063732	0.0160193	−0.40	0.691	N
Mission–Deploy	1.010632	0.0291142	0.0149732	1.94	0.052	M

SOURCE: RAND Arroyo Center analysis of TAPDB-AE and ReARMM data. N = 874.

Dimension 4: Match Between Soldiers' Preferred and Actual Duty and Location

This dimension analyzes the degree to which a soldier's assigned location matches the soldier's stated preference (when given), and whether that varies significantly over the ReARMM cycle. Table A.18 lists the variables considered in this analysis. An individual's record was deemed a match if that individual

- requested a base and was placed at that base (e.g., requested Aberdeen, got Aberdeen)
- requested a state and was placed in that state (e.g., requested Maryland, got Aberdeen)
- requested a base and was placed nearby (e.g., requested Bethesda, got Aberdeen)
- requested a state and was placed nearby (e.g., requested Delaware, got Aberdeen).

Table A.19 details the statistical results of this regression and notes whether the differences between other ReARMM phases and modernization are statistically significant.

TABLE A.18

TAPDB-AE Variables for MOS Match

Variable Name	TAPDB-AE Variable
Location Preference	CONAP
Assigned Location	LOCNM

SOURCE: RAND Arroyo Contor analysis of TAPDB-AF and ReARMM data.

TABLE A.19

Regression Statistics for ReARMM Phase and Preferred Location Match

ReARMM Phase	Mean	Delta	Standard Error	T	p>|t|	Significant?
Modernization	0.270963	--	--	--	--	--
Training	0.271988	0.0010247	0.0087606	0.12	0.907	N
Mission–PTDO	0.283325	0.0123617	0.0105717	1.17	0.243	N
Mission–Deploy	0.242320	−0.0286433	0.0098813	−2.90	0.004	Y

SOURCE: RAND Arroyo Center analysis of TAPDB-AE and ReARMM data. N = 874.

Dimension 5: Conflicts Between Education Attendance and ReARMM

We analyzed soldier attendance of education courses across each phase of the ReARMM cycle to see whether certain phases have greater levels of education attendance than others. In this analysis, "education attendance" refers to a soldier attending any resident course between 15 and 180 days.[1] We break the analysis into two categories: courses that last between 15 and 89 days ("shorter courses"), and courses that last between 90 and 180 days ("longer courses"). Tables A.20 and A.21 detail the statistical results of these regressions and note whether the differences between modernization and the other ReARMM phases are statistically significant.

---

[1] We do not include courses of less than 15 days or distance learning courses because of their lower disruption of ReARMM activities. Courses of more than 180 days are normally associated with a PCS, which we assess in Dimension 6.

TABLE A.20

Regression Statistics for ReARMM Phase and Shorter Course Attendance

ReARMM Phase	Mean	Delta	Standard Error	T	p>|t|	Significant?
Modernization	0.0336300	--	--	--	--	--
Training	0.0375362	0.0039100	0.0017514	2.23	0.026	Y
Mission–PTDO	0.0312941	−0.0023321	0.0021134	−1.10	0.270	N
Mission–Deploy	0.0237280	−0.0098982	0.0019754	−5.01	0.000	Y

SOURCE: RAND Arroyo Center analysis of TAPDB-AE, ATRRS, and ReARMM data. N = 894.

TABLE A.21

Regression Statistics for ReARMM Phase and Longer Course Attendance

ReARMM Phase	Mean	Delta	Standard Error	T	p>|t|	Significant?
Modernization	0.0053022	--	--	--	--	--
Training	0.0065299	0.0012277	0.0003395	3.62	0.000	Y
Mission–PTDO	0.0053482	0.000046	0.0004097	0.11	0.911	N
Mission–Deploy	0.0046588	−0.0006434	0.000383	−1.68	0.093	M

SOURCE: RAND Arroyo Center analysis of TAPDB-AE, ATRRS, and ReARMM data. N = 894.

Dimension 6: Conflicts Between Unit Turnover and ReARMM

This analysis measures the number of soldiers joining or leaving a unit (for example, due to an ETS or PCS) within a unit's ReARMM cycle to see whether there are measurable differences between modernization and the other phases. The analysis considered proportion of personnel gained by a unit in a given month, proportion of personnel lost, and the unit's growth ratio—i.e., how the overall number of individuals in a unit changed compared to the previous month. Tables A.22 and A.23 detail the statistical results of these regressions and note whether the differences between modernization and the other ReARMM phases are statistically significant.

TABLE A.22

Regression Statistics for ReARMM Phase and Unit Proportional Gains

ReARMM Phase	Mean	Delta	Standard Error	T	p>|t|	Significant?
Modernization	0.031284	--	--	--	--	--
Training	0.035897	0.0046137	0.0009552	4.83	0.000	Y
Mission–PTDO	0.032756	0.0014722	0.0011491	1.28	0.201	N
Mission–Deploy	0.028233	−0.003051	0.0010847	−2.81	0.005	Y

SOURCE: RAND Arroyo Center analysis of TAPDB-AE and ReARMM data. N = 812.

TABLE A.23

Regression Statistics for ReARMM Phase and Unit Proportional Losses

ReARMM Phase	Mean	Delta	Standard Error	T	p>|t|	Significant?
Modernization	0.032345	--	--	--	--	--
Training	0.035307	0.0029618	0.0009393	3.15	0.002	Y
Mission–PTDO	0.033935	0.0015900	0.0011299	1.41	0.160	N
Mission–Deploy	0.03124	−0.0011057	0.0010666	−1.04	0.300	N

SOURCE: RAND Arroyo Center analysis of TAPDB-AE and ReARMM data. N = 812.

Dimension 7: Associations Between Retention and People First Metrics

Dimension 7 explored the degree to which MOS match (Dimension 1) and assignment considerations (Dimension 2) are associated with retention. For this dimension, we analyzed the population of soldiers in FORSCOM brigades that had a reenlistment date between January 2020 and September 2022. Individuals who completed their term and reenlisted as of September 2022 (the end of our period of analysis) are categorized as "retained." Individuals who did not reenlist, or who separated prior to their ETS date, are categorized as "not retained."

To test associations with retention, we ran binary logistic regressions controlling for binary variables of gender, race, whether the soldier was assigned to a combat MOS, and soldier pay grade, as defined in Table A.24. To best isolate effects of the control variables, the model considered interaction terms between gender and the remaining variables (i.e., the model was fully interacted with gender).

This analysis investigated retention associations over seven variables of interest: first-degree MOS match, second-degree MOS match, third-degree MOS match, medical considerations, conduct considerations, standards considerations, and family considerations. Tables A.25 through A.32 detail the statistical results of the general model and the seven variables of interest.

TABLE A.24

Definitions of Binary Control Variables in Retention Regression Analysis

Control Variable	TAPDB-AE Variable	Value = 0	Value = 1
Gender	SEX	Female	Male
Race	RACPOP	White	Asian/Pacific Islander, Black, American Indian or Alaska Native, Other
Combat MOS	PMOSEN	All other MOS	MOS = 11, 12, 13, 15, 19
Pay Grade	PGRAD	E5, E6, E7, E8, E9	E1, E2, E3, E4

SOURCE: RAND Arroyo Center analysis of TAPDB-AE and ReARMM data.

TABLE A.25
Regression Statistics for General Retention Model

Variable	Risk Ratio	Standard Error	z	P>|z|	Significant?
Male	1.0617600	0.0159540	3.99	0.000	Y
Non-white	1.0825350	0.0145327	5.91	0.000	Y
Combat MOS	0.8998225	0.0192482	−4.93	0.000	Y
Lower Grade	0.9131790	0.0123171	−6.73	0.000	Y
Male*Non-white	0.9569963	0.0140060	−3.00	0.003	Y
Male*Lower grade	0.9492781	0.0138094	−3.58	0.000	Y
Male*Combat	1.0040160	0.0221768	0.18	0.856	N

SOURCE: RAND Arroyo Center analysis of TAPDB-AE and ReARMM data. N = 94,391.

TABLE A.26
Regression Statistics for Retention and First-Degree MOS Match

Variable	Risk Ratio	Standard Error	z	P>|z|	Significant?
MOS match	1.1354690	0.0937143	1.54	0.124	N
MOS*Male	1.1160100	0.0997178	1.23	0.219	N
Male	0.9530027	0.0853206	−0.54	0.591	N
Non-white	1.0820490	0.0145229	5.88	0.000	Y
Combat MOS	0.9015436	0.0192789	−4.85	0.000	Y
Lower Grade	0.9117991	0.0123159	−6.84	0.000	Y
Male*Non-white	0.9573009	0.0140061	−2.98	0.003	Y
Male*Lower grade	0.9486540	0.0138168	−3.62	0.000	Y
Male*Combat	1.0018130	0.0221206	0.08	0.935	N

SOURCE: RAND Arroyo Center analysis of TAPDB-AE and ReARMM data. N = 94,391.

TABLE A.27

Regression Statistics for Retention and Second-Degree MOS Match

Variable	Risk Ratio	Standard Error	z	P>|z|	Significant?
MOS/SQI match	1.0785010	0.0706341	1.15	0.249	N
MOS/SQI*Male	1.0483490	0.0715442	0.69	0.489	N
Male	1.0161570	0.0695500	0.23	0.815	N
Non-white	1.0822250	0.0145301	5.89	0.000	Y
Combat MOS	0.9020908	0.0193250	−4.81	0.000	Y
Lower Grade	0.9119430	0.0123419	−6.81	0.000	Y
Male*Non-white	0.9566096	0.0140002	−3.03	0.002	Y
Male*Lower grade	0.9459068	0.0138095	−3.81	0.000	Y
Male*Combat	1.0043810	0.0222176	0.20	0.843	N

SOURCE: RAND Arroyo Center analysis of TAPDB-AE and ReARMM data. N = 94,391.

TABLE A.28

Regression Statistics for Retention and Third-Degree MOS Match

Variable	Risk Ratio	Standard Error	z	P>|z|	Significant?
MOS/SQI/ASI match	1.0742180	0.0402556	1.91	0.056	M
MOS/SQI/ASI*Male	1.0606970	0.0415031	1.51	0.132	N
Male	1.0060610	0.0406461	0.15	0.881	N
Non-white	1.0819060	0.0145181	5.87	0.000	Y
Combat MOS	0.9055150	0.0194891	−4.61	0.000	Y
Lower Grade	0.9116300	0.0123256	−6.84	0.000	Y
Male*Non-white	0.9563281	0.0139874	−3.05	0.002	Y
Male*Lower grade	0.9466194	0.0137974	−3.76	0.000	Y
Male*Combat	1.0078740	0.0223994	0.35	0.724	N

SOURCE: RAND Arroyo Center analysis of TAPDB-AE and ReARMM data. N = 94,391.

TABLE A.29
Regression Statistics for Retention and Medical Considerations

Variable	Risk Ratio	Standard Error	z	P>|z|	Significant?
Medical	0.39919	0.015097	−24.28	0.000	Y
Medical*Male	1.22893	0.051287	4.94	0.000	Y
Male	1.02565	0.014442	1.80	0.072	M
Non-white	1.06224	0.013182	4.87	0.000	Y
Combat MOS	0.90654	0.018205	−4.89	0.000	Y
Lower Grade	0.94453	0.011830	−4.56	0.000	Y
Male*Non-white	0.97751	0.013320	−1.67	0.095	M
Male*Lower grade	0.91738	0.012464	−6.35	0.000	Y
Male*Combat	0.99680	0.020714	−0.15	0.880	N

SOURCE: RAND Arroyo Center analysis of TAPDB-AE and ReARMM data. N = 94,391.

TABLE A.30
Regression Statistics for Retention and Conduct Considerations

Variable	Risk Ratio	Standard Error	z	P>|z|	Significant?
Conduct	0.7079392	0.0174418	−14.02	0.000	Y
Conduct*Male	1.0348520	0.0272246	1.30	0.193	N
Male	1.0598680	0.0157716	3.91	0.000	Y
Non-white	1.0916820	0.0143503	6.67	0.000	Y
Combat MOS	0.9117320	0.0192055	−4.39	0.000	Y
Lower Grade	0.9459641	0.0125746	−4.18	0.000	Y
Male*Non-white	0.9659810	0.0138435	−2.42	0.016	Y
Male*Lower grade	0.9506399	0.0136399	−3.53	0.000	Y
Male*Combat	0.9981188	0.0217084	−0.09	0.931	N

SOURCE: RAND Arroyo Center analysis of TAPDB-AE and ReARMM data. N = 94,391.

TABLE A.31

Regression Statistics for Retention and Standards Considerations

Variable	Risk Ratio	Standard Error	z	P>|z|	Significant?
Standards	1.0300930	0.0228088	1.34	0.181	N
Standards*Male	0.9647081	0.0233683	−1.48	0.138	N
Male	1.0630090	0.0160412	4.05	0.000	Y
Non-white	1.0826060	0.0145291	5.91	0.000	Y
Combat MOS	0.8993180	0.0192365	−4.96	0.000	Y
Lower Grade	0.9107276	0.0123218	−6.91	0.000	Y
Male*Non-white	0.9568078	0.0140007	−3.02	0.003	Y
Male*Lower grade	0.9523200	0.0139007	−3.35	0.001	Y
Male*Combat	1.0044860	0.0221864	0.20	0.839	N

SOURCE: RAND Arroyo Center analysis of TAPDB-AE and ReARMM data. N = 94,391.

TABLE A.32

Regression Statistics for Retention and Family Considerations

Variable	Risk Ratio	Standard Error	z	P>|z|	Significant?
Family	0.7435434	0.0219126	−10.06	0.000	Y
Family*Male	1.4561530	0.0451280	12.13	0.000	Y
Male	1.0321190	0.0155149	2.10	0.035	Y
Non-white	1.0728050	0.0142502	5.29	0.000	Y
Combat MOS	0.8982038	0.0190653	−5.06	0.000	Y
Lower Grade	0.9247764	0.0123768	5.84	0.000	Y
Male*Non-white	0.9669531	0.0140241	−2.32	0.021	Y
Male*Lower grade	0.9368469	0.0135351	−4.52	0.000	Y
Male*Combat	1.0062570	0.0220638	0.28	0.776	N

SOURCE: RAND Arroyo Center analysis of TAPDB-AE and ReARMM data. N = 94,391.

Additional Background on the Scenario-Based Workshop and on the Expectation Management Literature

Appendix B provides additional details on two aspects of our research. In the first section, we provide descriptions of the tools we used for our scenario-based workshop. We also expand a little on the data collected. In the second section, we provide some highlights from our research on the topic of expectation management.

Scenario-Based Workshop

We conducted a two-hour workshop to assess Army leader priorities for balancing common Army tasks and People First initiatives. We assembled a convenience sample of field-grade Army officers to participate in role-based scenarios and surveys to explore how these officers worked through hypothetical friction points between People First initiatives and ReARMM requirements.

The day prior to the workshop, we emailed a survey to the officers to quantify a baseline understanding of how this cohort of officers prioritized People First initiatives. To understand potential impacts of the role-based scenario workshop on how officers viewed and prioritized People First implementation, we emailed a secondary survey immediately upon conclusion of the workshop. Officers were instructed to refrain from viewing or reflecting on their previous survey when completing the secondary survey. A copy of the survey is found in Figure B.1. The instructions provided to the officers are listed in Figure B.2.

FIGURE B.1

Example Survey Administered Immediately After the Workshop

	Option 1	Option 2	Option 3
Scenario 1	Civilian Schooling Requirement	MOS Training	Three day pass (earned/awarded)
	1	3	2
Scenario 2	Suicide Prevention Training	Special Mission Training (e.g., Regional orientation)	Routine Leave
	3	1	2
Scenario 3	Local Field Exercise	SHARP Training	Leave (Special Family Event)
	2	3	1
Scenario 4	Squad Leader Time/This is My Squad Time	Routine Leave	AR 350-1 Training (e.g. OPSEC)
	2	1	3
Scenario 5	Routine Leave	Weapons/Crew Ranges	Civilian Schooling Requirement
	3	2	1
Scenario 6	Major Training Exercise (NTC or MRX/MRE)	ETS Leave	Attend required PME course
	2	3	1
Scenario 7	On post military school (e.g. NBO)	Routine Leave	Unit METL Training (in Garrison)
	3	1	2

FIGURE B.2

Workshop Survey Instructions

1) Each set of boxes provides three options for you to prioritize/score, with "1" being highest priority and "3" being lowest. Place a 1, 2, and 3 under each option in the appropriate score row. No ties allowed, each row must have a 1, a 2, and a 3. Note that there are 7 total scenarios to evaluate.

2) For each set of three options, the question is, **as a Company Commander**, if a soldier asked you which option he/she will be allowed to do, **which would you tell them to do?**

For example, in the first case, **your unit** will be conducting a training exercise in the local training area and will be conducting SHARP Training over a time period when a soldier is requesting leave to attend a special family event. You need to rank order your preference for what the soldier will be allowed to do. If you rank Local Field Exercise "1", you are saying you would not approve the soldier's leave, because the exercise is more important. If you then rank Leave (Special Family Event) "2", you are saying that you would let the soldier miss SHARP training to attend the family event, but not ahead of the Local Field Exercise, etc.

In each case, the soldier can only do one of the three options. Which option do you prioritize, given the need to balance unit and soldier needs?

Officers were provided seven scenarios. In each scenario, they had three options, or events, to rank-order. The list of options the officers were asked to prioritize included the following three broad categories: time-off requests, non–Mission Essential Task List training, and Mission Essential Task List training. To gain insight into how these officers viewed work-life balance, which is a tenet of the People First initiative, one of the options always included some variation of a time-off request (e.g., routine leave, special pass, or leave in conjunction with separation from the service).

The two-hour workshop began with a brief discussion of the study's focus, provided information on nonattribution and voluntary basis of participation, and then began the role-playing scenarios. We crafted four scenarios, with four distinct roles per scenario, and allotted approximately 20 minutes for the officers to work through their role-playing problem set. We crafted each scenario to reflect a different type of Army brigade to capture varying viewpoints of the Army. The four scenarios are detailed in Figures B.3 through B.6.

FIGURE B.3

Workshop Garrison Support Scenario

Garrison Support Scenario

- Unit type: FA BDE

- Situation that is causing friction
  - Unit's phase in ReARMM: Training; this period of training is focused on the regional preparation for their next deployment.

  - Soldier/leader's dilemma: new Soldier needs to routinely arrive for work after first formation and do PT on their own because only Childcare they could find opens at 0830 (off post) and spouse doesn't drive/has physical limitations. On post CDC has 6-month wait list.

Player Motivations

- **Soldier**: Has no other reasonable solution for childcare; may be able to send family to parents for six months, but that would be a reason to leave Army instead.

- Company Commander/1SGT: sympathetic and supportive, but very concerned about other unit member's perceptions and reactions. Several other soldiers have difficult childcare issues.

- Garrison Staff – can't make exception for soldier, not fair to others on waiting list

- Spouse – trying to support soldier's career, but physically can't watch children all day. Doesn't drive (physical limitation).

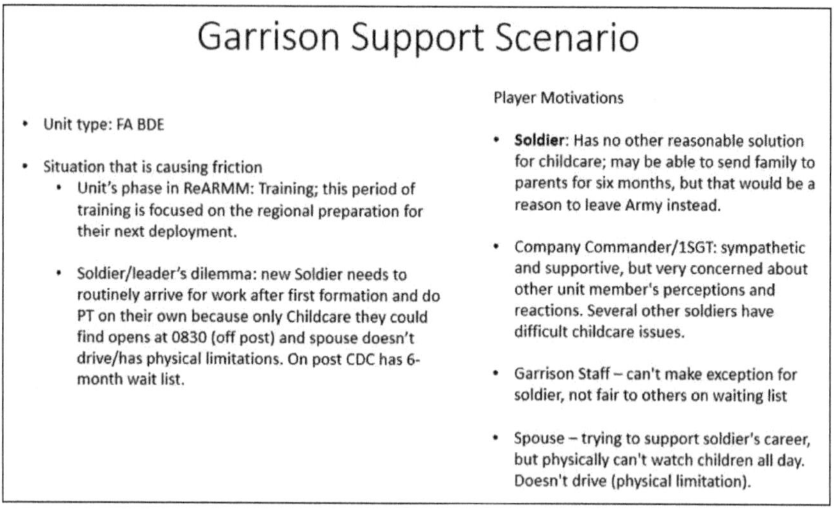

FIGURE B.4

Workshop Time-Off Scenario

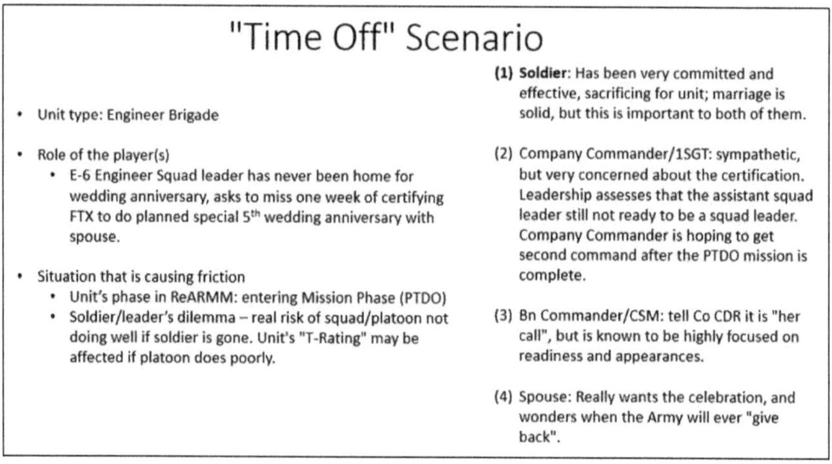

"Time Off" Scenario

- Unit type: Engineer Brigade

- Role of the player(s)
  - E-6 Engineer Squad leader has never been home for wedding anniversary, asks to miss one week of certifying FTX to do planned special 5th wedding anniversary with spouse.

- Situation that is causing friction
  - Unit's phase in ReARMM: entering Mission Phase (PTDO)
  - Soldier/leader's dilemma – real risk of squad/platoon not doing well if soldier is gone. Unit's "T-Rating" may be affected if platoon does poorly.

(1) **Soldier**: Has been very committed and effective, sacrificing for unit; marriage is solid, but this is important to both of them.

(2) Company Commander/1SGT: sympathetic, but very concerned about the certification. Leadership assesses that the assistant squad leader still not ready to be a squad leader. Company Commander is hoping to get second command after the PTDO mission is complete.

(3) Bn Commander/CSM: tell Co CDR it is "her call", but is known to be highly focused on readiness and appearances.

(4) Spouse: Really wants the celebration, and wonders when the Army will ever "give back".

FIGURE B.5

Workshop Armored Brigade Combat Team Fielding Scenario

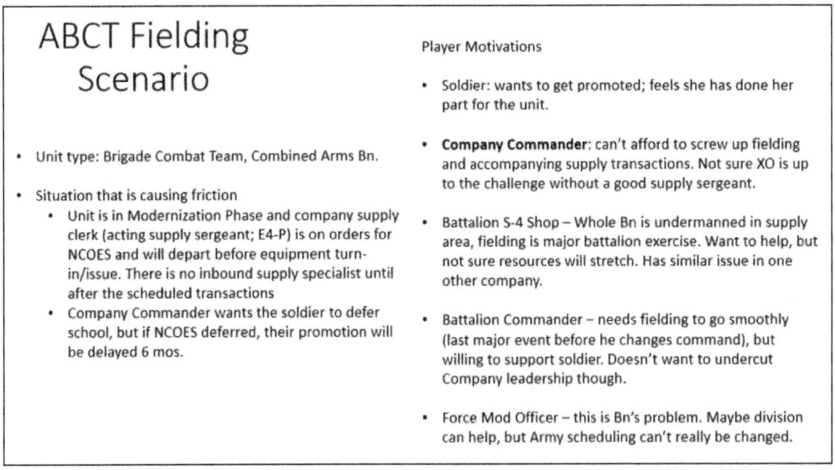

ABCT Fielding Scenario

- Unit type: Brigade Combat Team, Combined Arms Bn.

- Situation that is causing friction
  - Unit is in Modernization Phase and company supply clerk (acting supply sergeant; E4-P) is on orders for NCOES and will depart before equipment turn-in/issue. There is no inbound supply specialist until after the scheduled transactions
  - Company Commander wants the soldier to defer school, but if NCOES deferred, their promotion will be delayed 6 mos.

Player Motivations

- Soldier: wants to get promoted; feels she has done her part for the unit.

- **Company Commander**: can't afford to screw up fielding and accompanying supply transactions. Not sure XO is up to the challenge without a good supply sergeant.

- Battalion S-4 Shop – Whole Bn is undermanned in supply area, fielding is major battalion exercise. Want to help, but not sure resources will stretch. Has similar issue in one other company.

- Battalion Commander – needs fielding to go smoothly (last major event before he changes command), but willing to support soldier. Doesn't want to undercut Company leadership though.

- Force Mod Officer – this is Bn's problem. Maybe division can help, but Army scheduling can't really be changed.

FIGURE B.6

Workshop Support Tasking Scenario

Support Tasking Scenario

- Unit type: Med BDE

- Situation that is causing friction
  - Unit is in Training Phase of ReARMM
  - Unit is tasked to support Division's EFMB Testing
  - Supporting the testing will impact both general soldier skill and medical specialty training in the MED BDE

1) **Bn Commander**: I can't support this tasking, takes too many of my low-density MOSs away from our individual and collective training. I was ready to do this during modernization phase, but this is supposed to be our time to train.

2) **BDE S-3**: I can't tell Division "no", and the other Battalions haven't pushed back. I can't grow their tasking to save you. And you have a lot of "white space" on your training schedule, so how hard could this be?"

3) **Division Medical Officer**: EFMB is critically important to these soldiers. You'll have time to train later.

4) **Division G-3**: there is no "good time" to do this....no matter when we do EFMB someone will have to "pay" the support cost...figure it out....

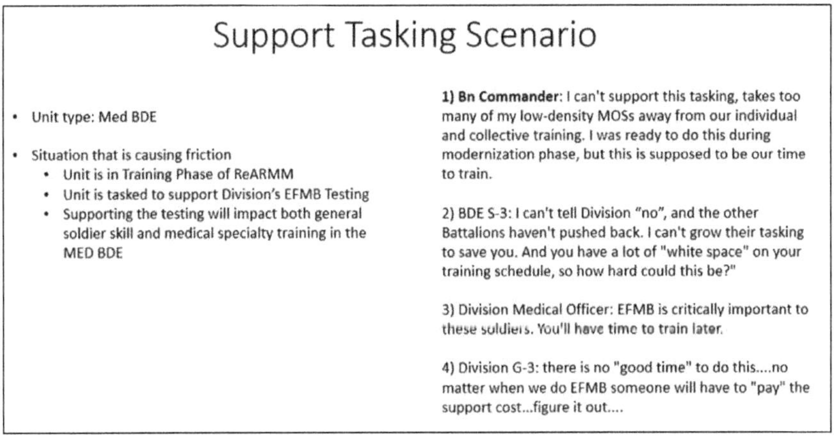

After the role-playing scenarios, the officers took a ten-minute break then reconvened to review the People First official guidance, receive a briefing on our study's preliminary findings, and allow for follow-on discussion. Immediately after the conclusion of the workshop, the officers completed the post-workshop survey.

Our analysis of the workshop surveys showed six instances, out of seven scenarios, in which a participant gave greater priority to granting a soldier leave after the role-playing than they did in the pre-workshop survey. And in the post-workshop survey, there were only two instances in which a participant gave lesser priority to a time-off (leave) request than they had in the pre-workshop survey.

This workshop provided interesting insight into how one cohort of field-grade officers prioritized People First initiatives when faced with a hypothetical conflict with ReARMM requirements.[1] This cohort demonstrated a noteworthy shift in how they rank-ordered People First initiatives immediately following a two-hour role-based scenario workshop, favoring increased priority to giving soldiers requested time off over mission requirements. This may be an interesting area of follow-on research to examine the following:

1. Would the changes in surveyed leaders' attitudes be statistically significant after a role-based or other People First initiative training platform in a sample size that is representative of the Army leaders' population?

2. Would the changes in leaders' attitudes in a hypothetical scenario translate to real-life changes in leadership practices?

3. How lasting would changes in leaders' attitudes be following training on People First initiatives?

4. If leaders' positive attitudes toward People First initiatives diminished over time, at what frequency would maintenance training possibly be of benefit? And, if maintenance training is of possible benefit, what method of delivery would best for maintenance training of People First initiatives?

---

[1] The scenario-based workshop sampled a small subset of field-grade officers. The small sample size was not representative of the officer or Army Total Force population and was not sufficiently powered to analyze with robust statistical methods.

Expectation Management

Expectation management can be defined as consistently communicating with key stakeholders (employees, bosses, clients, customer base) while shaping perceptions of one's intent, intended process, outcomes, and duration. In the context of the U.S. Army and the friction between People First and ReARMM, this translates into how leaders communicate with their soldiers not just about what will come next, but also about the level of change to expect in upcoming training schedules and unit mission focus.

The challenges facing Army leaders are significantly different than those facing civilian ones (or even government employees outside the military services). Nevertheless, the literature on expectation management and the associated communications demands is useful. As the main report highlighted, some of the greatest sources of friction we found came from inconstant messaging and from the unpredictability of a system designed to deliver predictability. We see these as shortfalls of expectation management, which include a lack of shared understanding of what to expect and when to expect it, closed communication channels, and unclear road mapping on the what and how of Army processes. The result is ambiguity and confusion among different working parts of the Army. Below we provide a very abbreviated set of observations from civilian literature on expectation management.

Expectation Management in a Civilian Construct

At a typical company, expectation management remains multifaceted; expectations are managed across an array of stakeholders, managers, and team members. The civilian workforce provides a dynamic relationship in the workforce requiring training and incentives for employees of companies in both public and private sectors. Civilian leaders may have the flexibility for personal circumstance, shown in the way a salary worker is able to use sick leave, paid time off, or other benefits more freely than service members. The expectation management practices regarding goals and outcomes may also be inherently different in the civilian world, as companies rely on signals from the marketplace (both for labor and for their product) that are inherently different than the types of feedback military organizations get.

For example, a typical company business model is defined by meeting both customer and workplace expectations.[2] If the goal is to have happy customers—those whose expectations of a product or service have been met or exceeded—then employers communicate relevant expectations to employees to ensure the quality of a product or service. In this way, employees understand where their priorities lie and focus on the correct tasks in the order of greater importance as instructed by management. On the flip side, employers and managers must balance employee needs and satisfaction to create a sustainable and productive working environment.

In an opinion piece by a senior director at Tubi, "Expectation Management: A Manager's Guide," Blake Bassett offers a guide on managing expectations in a company setting.[3] They can be summarized as:

- **"Publish your expectations":** Post the expectations for the team in a place where everyone will read them. Expectations should translate the company's expectations into team-specific guidance with concrete examples. Additionally, managers are encouraged to post team expectations as well as team expectations of the manager.
- **"Establish a North Star"** (vision): Set an ideal end state for a project, initiative, organization, or product as a means to enable employees to navigate ambiguity and "make decisions and tradeoffs in the absence of direct guidance."
- **"Understand the how":** Have managers provide guidance during goal-setting to establish a mutual understanding of what is to be achieved and the "how" behind the goal.
- **"Seek mutual understanding with stakeholders":** Emphasize effective communication with external parties pertaining to scope of responsibilities. Also consider creating a statement of work outlining the problems that the team plans to solve, objectives, what success looks like, measurements of success, and the steps to take to accomplish objec-

---

[2]   Daniel Decker, "Expectation Management: The Secret to Happy Customers and Rapid Growth," *Forbes*, September 19, 2018.

[3]   Blake Bassett, "Expectation Management: A Manager's Guide," *Medium*, February 3, 2020.

tives. Most importantly, this statement of work should identify out-of-scope work.

- **"Reinforce and adjust"**: Within the management cycle, reinforce expectations and make necessary adjustments to keep the team on track by gathering updates of members' progress, reiterating guidance, and providing clear and concrete feedback.

This summary echoes various pieces written on managing expectations within a leader-team dynamic, ranging from the Forbes Coaches Council expounding on the importance of setting attainable expectations to papers on management modeling, showing that this thought holds true across industry.[4]

In a field manual on software development, authors Raphael Malveau and Thomas J. Mowbray articulate the importance of managing expectations as stakes increase and impact how work is received or acknowledged by different internal and external groups (Malveau and Mowbray, 2003). Described as a powerful weapon in psychological warfare, there are three key elements to ensure success. First, by overemphasizing the negatives of an idea, dissatisfaction and a loss of confidence in the promiser's ability are kept to a minimum. At the same time, should the consequence intended be achieved, then satisfaction is guaranteed because more was delivered than expected. Second, Malveau and Mowbray suggest the tactic of underpromising to overdeliver. By explaining the caveats of an achievable goal, a psychological framework of expectations is established, and workers may be able to deliver upon expectations. Without this framework, the stakeholder might perceive the promiser to be underperforming—even when achieving the same end result—due to poorly managed perception engineering. Last, in order for the first two elements to hold true, clear, back-and-forth communication channels must be established so that the manager can communicate what, how, and when to the team. Additionally, they must also communicate risks and information updates to stakeholders. When all three are accomplished, management flows smoothly due to a constant source of feedback from all parties, low morale is minimized, and deliverables can be

---

[4] Forbes Coaches Council, "13 'Right' Ways for Leaders to Set Expectations with Employees," Forbes, November 24, 2021.

received in a more favorable light even when they did not meet the planned outcome. In this way, businesses are able to adapt their personnel systems to meet changing requirements.[5]

The efficacy of these theoretical approaches can be demonstrated by a case study of German public perception, salience, and effectiveness of regulatory measures during the early phase of the pandemic.[6] This study found that reinforcing and repeating a message allows the public to remember and be more cautious, which led to people wearing masks and success tackling the pandemic. Additionally, former German Chancellor Angela Merkel's method of planning for the long term, avoiding articulating incomplete information and communicating risks clearly, allowed her to minimize public discontent and shift citizens' expectations about the duration of the pandemic. The last takeaway can be that communicators can use the momentum at the beginning of a major change before effectiveness begins to deteriorate over time.

In the same vein, Army leadership impacts unit and soldier expectations of their chain of command—a sentiment echoed by Everett S. P. Spain's article "Managing Expectations While Leading Change."[7] Spain provides a similar definition on managing expectations, in which

> [o]ne consistently communicated with key stakeholders to understand their spoken and unspoken expectations, while realistically shaping their perceptions of your true character and intentions, the benefits of the long-term change process, what constitutes short-term success, and that stakeholder's specific responsibilities required to achieve the short and long-term outcomes.[8]

---

[5] J. D. Meier, "Expectation Management," *Sources of Insight*, December 1, 2007.

[6] Peter Haan, Andreas Peichl, Annekatrin Schrenker, Georg Weizsäcker, and Joachim Winter, "Expectation Management of Policy Leaders: Evidence from COVID-19," *Journal of Public Economics*, May 2022.

[7] Everett S. P. Spain, "Managing Expectations While Leading Change," *Military Review*, Vol. 87, March–April 2007.

[8] Spain, 2007, p. 75.

This is consistent with past literature and interview analysis and solidi-fies the importance of open communication channels. Spain also provides 12 key lessons learned from his time as a company commander.

1. Underpromise and overdeliver.
2. Set short-term goals with your key stakeholders.
3. Have your stakeholders commit in a public setting.
4. Use message repetition to communicate clarity.
5. Changing the message is a strength, not a weakness.
6. Set up regular meetings and a communication center.
7. Managing expectations calls for establishing two-way communica-tion.
8. Always communicate what is not possible and why.
9. The organizational leader should lead the managing expectations efforts.
10. Being positive is a catalyst in managing expectations.
11. Don't fear inevitable incidents; just respond promptly to them.
12. Get around egos by always using a full spectrum of communication.

The sum of these lessons provides an in-depth look at the relation-ship between military superiors and their relationship with subordinates as change agents, with much of the writing addressing the gap in commu-nication, reasoning, and information understanding felt by servicemem-bers in the interviews. Interestingly, these lessons similarly reflect investors and their expectations regarding investment decisions in the stock market. Here, it is important to note that market prediction and risk are synony-mous. A successful ex ante of stock price analyses for market prediction allows management to estimate the company's future performance compar-ing the expectations implied by the current stock price with its own expecta-tions, "identifying the shortfall in its corporate plan, and then ferreting out restructuring opportunities to minimize the shortfall."[9] Conversely, should the reverse be true, then management would have an opportunity to com-municate information back to the market in order to raise its expectations.

---

[9] Alfred Rappaport, "Stock Market Signals to Managers," *Harvard Business Review*, Vol. 65, No. 6, November 1987.

Through this, stock pricing becomes a measurement of market performance expectations. In the same vein, a company commander's ability to communicate "good" expectation management impacts units' attitudes and understanding of their purpose and the Army's objectives. Additionally, unit leaders are ideally able to gauge soldiers' expectations and begin to adjust their trajectory accordingly—or minimize shortfalls of set expectations. By being able to generate predictions of work performance for both parties, senior leadership would be able to increase transparency in the decisions made and communicate clearly.

Conclusion

This appendix provided additional details on our scenario-based workshop. Our key takeaway from that tool was that there may be an important role in Army training and leader development for similar kinds of exercises. This type of scenario-based structured discussion could shape how Army leaders perceive and react to leadership dilemmas in the absence of clear guidance and rubrics.

On the topic of expectation management, we note the unique challenges of managing uncertainty. Our expectation is not that the Army adopt the recommendations or concepts we provide as a whole. Rather, we provide some perspectives and tools that the Army could incorporate as it thinks through how to provide (what we believe to be necessary) training to its leaders on how to best communicate about uncertainty with its soldiers.

Abbreviations

ABCT	armored brigade combat team
ARFORGEN	Army Force Generation
ASI	Additional Skill Identifier
ATRRS	Army Training Requirements and Resources System
BCT	brigade combat team
CSA	Chief of Staff of the Army
CTC	combat training center
DOTMLPF-P	doctrine, organization, training, materiel, leadership, policy, facilities, personnel
ETS	expiration of time in service
FOC	full operating capability
FORSCOM	U.S. Army Forces Command
FY	fiscal year
HQDA	Headquarters, Department of the Army
HRC	Human Resources Command
IOC	initial operating capability
MOS	military occupational specialty
MTOE	modified table of organization and equipment
NCO	noncommissioned officer
NCOER	noncommissioned officer efficiency reporting
NET/NEF	new equipment training/new equipment fielding
OER	officer efficiency reporting
OPTEMPO	operational tempo
PCS	permanent change of station
PME	professional military education
PTDO	prepare to deploy order
RA	Regular Army
RC	reserve component
ReARMM	Regionally Aligned Readiness and Modernization Model

SHARP   Sexual Harassment/Assault Response and Prevention
SQI   Special Qualification Identifier
SRM   Sustainable Readiness Model
TAPDB-AE   Total Army Personnel Database for Army Enlisted
TRADOC   U.S. Army Training and Doctrine Command

Bibliography

82nd Airborne Division Annual Training Guidance, FYs 2022–2023.

101st Airborne Division Annual Training Guidance, FYs 2022–2023.

Army Doctrine Publication 6-0, *Mission Command, Command and Control of Army Forces*, Headquarters, Department of the Army, July 2019.

Army Doctrine Publication 6-22, *Army Leadership and the Profession*, Headquarters, Department of the Army, July 2019.

Army Regulation 220-1, *Army Unit Status Reporting and Force Registration—Consolidated Policies*, Headquarters, Department of the Army, August 16, 2022.

Army Regulation 525-29, *Army Force Generation*, Headquarters, Department of the Army, March 14, 2011.

Army Regulation 525-29, *Force Generation—Sustainable Readiness*, Headquarters, Department of the Army, October 1, 2019.

Army Regulation 600-20, *Personnel—General Army Command Policy*, Headquarters, Department of the Army, July 24, 2020.

Bannister, Alec, "Bravo Battery, 3-4 ADA BN 108th ADA BDE," *Spartan Magazine*, undated.

Bassett, Blake, "Expectation Management: A Manager's Guide," *Medium*, February 3, 2020.

Bates, Latashia, *Army Readiness and Modernization in 2022*, Association of the United States Army, Land Warfare Papers, No. 146, 2022.

Blue Star Families, *2021 Military Family Lifestyle Survey Comprehensive Report*, 2022.

Bonds, Timothy M., Dave Baiocchi, and Laurie L. McDonald, *Army Deployments to OIF and OEF*, RAND Corporation, DB-587-A, 2010. As of September 14, 2023:
https://www.rand.org/pubs/documented_briefings/DB587.html

Britzky, Haley, "Soldiers Say the Army's Relentless Push for Readiness Is 'Breaking the Force' in Leaked Documents," *Task & Purpose*, September 20, 2019.

Britzky, Haley, "Army Chief of Staff Gen. James McConville Wants Soldiers to Have Work-Life Balance," *Task & Purpose*, October 15, 2020.

Britzky, Haley, "The Sgt. Maj. of the Army Wants Leaders to Stop Scheduling Training Just for the Sake of It," *Task & Purpose*, October 18, 2021.

Buddin, Richard, *Success of First-Term Soldiers: The Effects of Recruiting Practices and Recruit Characteristics*, RAND Corporation, MG-262-A, 2005. As of January 11, 2023:
https://www.rand.org/pubs/monographs/MG262.html

Butler, Dwayne M., Angelena Bohman, Lisa Pelled Colabella, Julia A. Thompson, Michael Shurkin, Stephan B. Seabrook, Rebecca Jensen, and Christina Bartol Burnett, *Comprehensive Analysis of Strategic Force Generation Challenges in the Australian Army*, RAND Corporation, RR-2382-AUS, 2018. As of September 12, 2023:
https://www.rand.org/pubs/research_reports/RR2382.html

Campbell, Charles C., "ARFORGEN: Maturing the Model, Refining the Process," *Army Magazine*, Vol. 59, No. 6, June 2009.

Decker, Daniel, "Expectation Management: The Secret to Happy Customers and Rapid Growth," Forbes, September 19, 2018.

Feickert, Andrew, *The Army's Regionally Aligned Readiness and Modernization Model*, Congressional Research Service, IF11670, Version 3, September 22, 2022a.

Feickert, Andrew, *Defense Primer: Army Multi-Domain Operations (MDO)*, Congressional Research Service, IF11409, November 21, 2022b.

Feickert, Andrew, and Lawrence Kapp, *Army Active Component (AC)/ Reserve Component (RC) Force Mix: Considerations and Options for Congress*, Congressional Research Service, R43808, 2014.

Field Manual 6-22, *Developing Leaders*, Headquarters, Department of the Army, November 2022.

Field Manual 7-0, *Training*, Headquarters, Department of the Army, June 2021.

Forbes Coaches Council, "13 'Right' Ways for Leaders to Set Expectations with Employees," Forbes, November 24, 2021.

FORSCOM—*See* U.S. Army Forces Command.

Garraton, Ricardo R., *Analysis of Army Force Generation Model Behavior and Expectation Management*, U.S. Army War College, December 3, 2012.

Haan, Peter, Andreas Peichl, Annekatrin Schrenker, Georg Weizsäcker, and Joachim Winter, "Expectation Management of Policy Leaders: Evidence from COVID-19," *Journal of Public Economics*, May 2022.

Headquarters, Department of the Army, Deputy Chief of Staff, G-3-5-7, "Regionally Aligned Readiness and Modernization Model," webpage, October 16, 2020. As of September 14, 2023:
https://www.army.mil/standto/archive/2020/10/16/

Helmus, Todd C., S. Rebecca Zimmerman, Marek N. Posard, Jasmine L. Wheeler, Cordaye Ogletree, Quinton Stroud, and Margaret C. Harrell, *Life as a Private: A Study of the Motivations and Experiences of Junior Enlisted Personnel in the U.S. Army*, RAND Corporation, RR-2252-A, 2018. As of September 14, 2023:
https://www.rand.org/pubs/research_reports/RR2252.html

Hemmerly-Brown, Alexandra, "ARFORGEN: Army's Deployment Cycle Aims for Predictability," *STAND-TO!*, November 19, 2009.

Institute for Operations Research and the Management Sciences, "Operations Research and Analytics," webpage, undated. As of February 8, 2023:
https://www.informs.org/Explore/Operations-Research-Analytics

Markel, M. Wade, Alexandra T. Evans, Miranda Priebe, Adam Givens, Jameson Karns, and Gian Gentile, *The Evolution of U.S. Military Policy from the Constitution to the Present, Volume IV: The Total Force Policy Era, 1970–2015*, RAND Corporation, RR-1995/4-A, 2020. As of September 14, 2023:
https://www.rand.org/pubs/research_reports/RR1995z4.html

Marrone, James V., S. Rebecca Zimmerman, Louay Constant, Marek N. Posard, Katherine L. Kidder, Christina Panis, and Rebecca Jensen, *Organizational and Cultural Causes of Army First-Term Attrition*, RAND Corporation, RR-A666-1, 2021. As of September 12, 2023:
https://www.rand.org/pubs/research_reports/RRA666-1.html

Malveau, Raphael C., and Thomas J. Mowbray, *Software Architect Bootcamp*, 2nd ed., Prentice Hall, December 10, 2003.

McConville, James C., "People First: Insights from the Army's Chief of Staff," *Army Sustainment*, Vol. 53, No. 1, 2021.

McHugh, John M., and George W. Casey, *America's Army: The Strength of the Nation at a Strategic Crossroads—A Statement on the Posture of the United States Army, Fiscal Year 2012*, posture statement presented to the 112th Congress, 1st Session, U.S. Department of the Army, 2011, Addendum F.

Meier, J. D., "Expectation Management," *Sources of Insight*, December 1, 2007.

Milley, Mark A., "Memorandum for All Army Leaders, Subject: Army Readiness Guidance, Calendar Year 2016–7," Chief of Staff, Army, January 20, 2016.

Orvis, Bruce R., Christopher E. Maerzluft, Sung-Bou Kim, Michael G. Shanley, and Heather Krull, *Prospective Outcome Assessment for Alternative Recruit Selection Policies*, RAND Corporation, RR-2267-A, 2018. As of January 11, 2023:
https://www.rand.org/pubs/research_reports/RR2267.html

Rappaport, Alfred, "Stock Market Signals to Managers," *Harvard Business Review*, Vol. 65, No. 6, November 1987.

Reinsch, Michael, "People First Task Force Building More Cohesive Teams," Army News Service, December 13, 2021.

Ryan, Kurt J., and Jin H. Pak, "Operationalizing ReARMM: A Sustainment Perspective," *Army Sustainment*, August 11, 2021.

Saum-Manning, Lisa, Tracy C. Krueger, Matthew W. Lewis, Erin N. Leidy, Tetsuhiro Yamada, Rick Eden, Andrew Lewis, Ada L. Cotto, Ryan Haberman, Robert Dion, Jr., Stacy L. Moore, Michael Robert Shurkin, and Michael Lerario, *Reducing the Time Burdens of Army Company Leaders*, RAND Corporation, RR-2979-A, 2019. As of September 21, 2023: https://www.rand.org/pubs/research_reports/RR2979.html

SocioCultural Research Consultants, Dedoose, web application for managing, analyzing, and presenting qualitative and mixed-method research data, version 9.0.84, 2023.

Spain, Everett S. P., "Managing Expectations While Leading Change," *Military Review*, Vol. 87, March–April 2007.

Spilka, Dymtro, "3 Ways to Manage Expectations When Starting Out as an Investor," Credit.com, February 8, 2022. As of February 3, 2023: https://www.credit.com/blog/ways-to-manage-expectations-investor/

Suits, Devon, "Army Implementing ReARMM Unit Life Cycle Model" Army News Service, March 2, 2021.

Swecker, Christopher, Jonathan Harmon, Carrie Ricci, Queta Rodriquez, and Jack White, *Report of the Fort Hood Independent Review Committee*, U.S. Army, November 6, 2020. As of September 21, 2023: https://www.army.mil/e2/downloads/rv7/forthoodreview/2020-12-03_FHIRC_report_redacted.pdf

Tan, Michelle, "Putting People First: McConville Looks to Revolutionize How Soldiers Serve," Association of the United States Army, October 3, 2019.

U.S. Army, "Army People First: Prioritizing Our Most Valuable Asset—People First Task Force," webpage, undated. As of February 8, 2023: https://www.army.mil/peoplefirst/#task-force

U.S. Army, *U.S. Army Talent Management Strategy: Force 2025 and Beyond— Ready, Professional, Diverse, and Integrated*, Headquarters, Department of the Army, September 20, 2016.

U.S. Army, "Army People Strategy: Overview," webpage, 2019a. As of February 7, 2023: https://people.army.mil/overview-2/

U.S. Army, *The Army People Strategy*, October 2019b.

U.S. Army, "Secretary of the Army Announces Missing Soldier Policy, Forms People First Task Force to Implement Fort Hood Independent Review Committee (FHIRC) Recommendations," webpage, December 8, 2020. As of February 8, 2023:
https://www.army.mil/article/241490

U.S. Army Forces Command, "Army Force Generation," *STAND-TO!*, July 19, 2010. As of September 19, 2023:
https://www.army.mil/article/42519/army_force_generation

U.S. Army Public Affairs, "Army Announces CID Restructure and SHARP Policy Improvements," webpage, May 6, 2021. As of February 8, 2023:
https://www.army.mil/article/246054

U.S. Army Talent Management, homepage, undated-a. As of September 23, 2023:
https://talent.army.mil/

U.S. Army Talent Management, "Army Talent Alignment Process," webpage, undated-b. As of September 23, 2023:
https://talent.army.mil/atap/

U.S. Army War College, *How the Army Runs: A Senior Leader Handbook, 2019–2020*, 2020.

U.S. Government Accountability Office, *Army Readiness: Progress and Challenges in Rebuilding Personnel, Equipping, and Training*, testimony before the Subcommittee on Readiness and Management Support, Committee on Armed Services, U.S. Senate, statement of John H. Pendleton, director, Defense Capabilities and Management, GAO-19-367T, February 6, 2019.

Wong, Leonard, and Stephen J. Gerras, *Lying to Ourselves: Dishonesty in the Army Profession*, U.S. Army War College Strategic Studies Institute, 2015.

Wormuth, Christine E., and James P. McConville, "Statement by the Honorable Christine E. Wormuth, Secretary of the Army, and General James P. McConville, Chief of Staff United States Army, Before the Committee on Armed Services, United States Senate, First Session, 117th Congress, on the Posture of the United States Army," U.S. Senate, June 15, 2021.

Zimmerman, S. Rebecca, Kimberly Jackson, Natasha Lander, Colin Roberts, Dan Madden, and Rebeca Orrie, *Movement and Maneuver: Culture and the Competition for Influence Among the U.S. Military Services*, RR-2270-OSD, RAND Corporation, 2019. As of September 24, 2023:
https://www.rand.org/pubs/research_reports/RR2270.html

Milton Keynes UK
Ingram Content Group UK Ltd.
UKHW050416160224
437760UK00007B/24